Study Guide

**Volumes
2 and 3**

Sears and
Zemansky's

UNIVERSITY
PHYSICS

**with Modern Physics
11th Edition**

Young & Freedman

JAMES R. GAINES
University of Hawaii

WILLIAM F. PALMER
Ohio State University

PEARSON

Addison
Wesley

San Francisco • Boston • New York
Capetown • Hong Kong • London • Madrid • Mexico City
Montreal • Munich • Paris • Singapore • Sydney • Tokyo • Toronto

ISBN 0-8053-8743-9

1 2 3 4 5 6 7 8 9 10–PBT–07 06 05 04 03
www.aw.com/bc

CONTENTS

PREFACE

The emphasis in any calculus-based physics course for scientists and engineers is on problem solving. This Guide as been written to help you learn how to solve the kind of problems you will encounter in homework assignments and examinations.

Each chapter in the Guide corresponds to a chapter in the parent text, *University Physics*, Eleventh Edition, by Young and Freedman (Addison Wesley, 2004). The organization of the chapters is as follows:

Objectives. All chapters begin with a statement of learning objectives, stressing application of basic physical principles to problems and calculations.

Review. Primary concepts, definitions, formulas, units, and physical constants are then reviewed and highlighted.

Supplement. Some chapters have supplementary material which builds the basis for or expands upon materials of the text. These supplements are primarily in the early chapters, where they are concerned with applying basic skills of vector manipulation and calculus to physical problems.

Problem-Solving Strategies. Many chapters have a section on problem solving strategies, including step-by-step guidance on how to set up and complete a problem.

Questions and Answers. This section poses questions that are conceptual rather than mathematical in nature. Occasionally, a short calculation is required to answer a question.

Examples and Solutions. In this section examples similar to the problems at the end of the chapter in the text are worked out, step-by-step, with commentary on organization of given information, principles applied, set-up procedure, pitfalls, and alternative solutions.

Quiz. Each chapter has a short quiz, with answers. In addition, any of the solved examples may be used as a self-quiz.

The Guide is thus an organizer of study, review, and self-diagnostic activities, as well as a ready reference of solved problems. If difficulty is encountered with a homework problem, a similar one can usually be found in the Guide, fully worked out. Rather than stare at the homework problem, with no useful forward progress, you can refer to the similar one for guidance. This is one of the chief uses to which the Guide is put. Solved exercises as a pedagogical tool are a tradition in physics.

We have attempted to make the solutions clear and detailed. We have emphasized problems which demonstrate the main concepts, avoiding those which require only substitution into a formula of the text.

After reading the text and notes you may wish to read the objective and review sections, and then work through the examples, referring back to the text and supplement sections when an unfamiliar concept or method is met. This process may occur before, after, or during the process of attempting the homework.

A word of caution is in order when reading through solved problems. The acid test of understanding comes only when you challenge yourself by taking a self-administered examination. This occurs each time you sit down to do a homework problem, or you can make things more realistic by testing yourself with an example in the Guide under examination conditions: Do not peek at the solution. Use no books or notes. Then score yourself according to correct method and correct answer.

The results of such an exercise may be enlightening. Remember your instructor regards the statement "I understand the material but don't know how to solve the problems" as self-contradictory. In a sense, the problems **are** the material. A complete mastery of the material consists in the ability to (I) <u>identify</u> the appropriate physics concepts underlying the problem, (2) <u>set up</u> the problem, applying these concepts, (3) <u>execute</u> a solution with confidence and care to a final result, and (4) <u>evaluate</u> your result to ensure that it makes sense and that there are no computational errors.

To that end we hope this Study Guide will make your way a little less rough. Good luck!

James R. Gaines
William F. Palmer
Honolulu and Columbus, 2003

21
ELECTRIC CHARGE AND ELECTRIC FIELD

OBJECTIVES

In this chapter you will be introduced to *electric charge* and the *Coulomb force* between *positive* and *negative* charges. Insulating and conducting qualities are discussed. *Atoms* will be described as made up of negatively charged *electrons* and a *nucleus* made up of positively charged *protons* and neutral *neutrons*, leading to the ideas of charge conservation and quantization. The *electric field*, a *vector*, is introduced. The field strength at a point is calculated by direct summation of the fields due to all other charges. Your objectives are to:

> Learn how bodies become *charged*, positively and negatively, by physical processes involving charge separation.

> Observe that like charges repel; unlike attract.

> Identify charged bodies as having an *electron surplus or deficit*.

> Charge an electroscope by physical charge *transfer or induction*.

> Calculate the *force between point charges,* using Coulomb's law.

> Calculate the *electric field* caused by a distribution of point charges by vector summation of the electric fields of the individual point charges.

> Calculate the electric field of *continuous charge distributions* by summing the field of its parts. Examples are a line charge, an infinite plane, or the axial field of a circular loop of charge.

REVIEW

When two dissimilar materials are rubbed together, one material often becomes *negatively charged* and the other *positively charged*. The materials are originally *neutral* (uncharged), containing a charge balance between *negatively charged electrons* and the *positively charged nucleus*. When rubbed, the negative electrons accumulate on one material and are removed from the other. The charge is said to have been separated. The material with a surplus of electrons is *negatively charged*. The material with a deficit of electrons is *positively charged*. Bodies with like charges repel each other. Bodies with unlike charges attract each other.

The magnitude of the force between two charged objects is given by Coulomb's law,

$$F = \frac{k[qq']}{r^2} = \frac{1}{4\pi\varepsilon_0} \frac{[qq']}{r^2}$$

$$k = \frac{1}{4\pi\varepsilon_0} = 8.99 \times 10^9 \text{ Nm}^2/\text{C}^2$$

$$\varepsilon_0 = 8.85 \times 10^{-12} \text{ C}^2/\text{N}\cdot\text{m}^2$$

yielding F in newtons (N) when q and q' are in *coulombs* (C) and r is in meters (m). The coulomb is a new, independent unit, not previously encountered. It has an independent status, as do m, kg, s, in the SI system.

Matter is generally made up of neutral atoms. A *neutral* atom consists of Z negatively charged electrons surrounding a positively charged *nucleus* containing Z positively charged *protons* and A-Z *neutrons*. Z is the atomic number of the nucleus.

The charge of an electron is the negative number -e, where

$$e = 1.60 \times 10^{-19} \text{ C}.$$

The mass of an electron is

$$\text{Mass of electron} = 9.10 \times 10^{-31} \text{ kg}.$$

The charge on a proton is e, and the mass of a proton is

$$\text{Mass of proton} = 1.67 \times 10^{-27} \text{ kg} = 1.67 \times 10^{-24} \text{ g}$$

$$= \text{about 1800 electron masses.}$$

The mass of the neutron is approximately that of the proton,

$$\text{Mass of neutron} = 1.67 \times 10^{-27} \text{ kg}.$$

The charge of the neutron is zero.

The atomic number Z of an atom is the number of protons in its nucleus. A gram-atom or mole of a pure substance is an amount of the substance equal in grams to its atomic weight. The mass of a gram-atom of hydrogen is primarily due to the protons within it, with small corrections for the electrons and binding energies. Thus the number of protons in a gram-atom of hydrogen is

$$\frac{(1.008 \text{ g})}{\left(1.673 \times 10^{-24} \text{ g}\right)} = 6.02 \times 10^{23} = \text{Avogadro's number.}$$

An electric field E exists at a point in space if there is a force F of electric origin on a test charge q' at rest at that point.

By "electric origin" is meant that the force, and electric field, is due to all other charged bodies present. Since the test charge itself may disturb the surrounding charges, it is desirable to minimize this effect by making q' as small as possible. The electric field $\vec{E}$ is defined as

$$\vec{E} = \text{Lim}_{q' \to 0} \left(\frac{\vec{F}}{q'} \right)$$

where $\vec{F}$ is the Coulomb force on charge q'.

The electric field points in the direction that a positive test charge would move if placed in the field at rest. The static field lines of force are continuous in charge free regions; for static distributions of charges they begin on positive charges and end on negative charges.

If the force F is in newtons (N) and q' is in coulombs (C) the field E has units of N/C.

The field due to a point charge of magnitude q is given by

$$\vec{E} = \frac{1}{4\pi\varepsilon_0} \frac{q}{r^2} \hat{r} = k\frac{q}{r^2} \hat{r}$$

where $\hat{r}$ is a unit vector pointing radially outward from the charge at the field point P (along $\hat{r}$). See Fig. 21-1. If the charge q is positive, $\vec{E}$ has the direction of $\hat{r}$. If the charge is negative, $\vec{E}$ has the opposite direction.

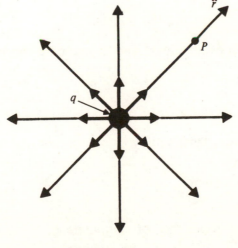

Figure 21-1

The field strength is inversely proportional to the square of the distance from P to q; the field strength is proportional to the number of field lines per unit area perpendicular to the field direction. It is greatest near the charge q where the field lines are closely spaced.

3

The field of a set of point charges q_i is the *vector sum* (or "superposition") of the fields of the individual point charges,

$$\mathbf{E} = \sum_i \frac{1}{4\pi\varepsilon_0} \frac{q_i \hat{\mathbf{r}}_i}{r_i^2}$$

If the charge distribution is continuous, the sum becomes an integral

$$\vec{E} = \sum_i \frac{1}{4\pi\varepsilon_0} \frac{q_i \hat{r}_i}{r_i^2} \rightarrow \frac{1}{4\pi\varepsilon_0} \int \frac{(dq)\hat{r}}{r^2}$$

where $\hat{r}$ points from the charge increment dq to the field point as in Fig. 21-1.

The axial field of a charge Q uniformly distributed on a ring of radius a is

$$E_x = \frac{1}{4\pi\varepsilon_0} \frac{Qx}{(x^2 + a^2)^{3/2}} \, ,$$

as shown in Fig. 21-2. (Components not in the axial (x) direction cancel, as shown.)

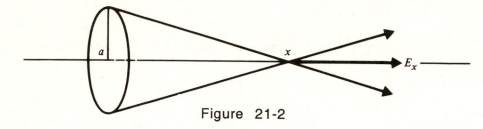

Figure 21-2

As $a \rightarrow 0$ the loop behaves like a point charge, and E_x approaches the field due to a point charge,

$$E_x \cong \frac{1}{4\pi\varepsilon_0} \frac{Q}{x^2} .$$

The field of a long line charge described by its charge per unit length (λ)

$$\lambda = \frac{dQ}{dL} = \text{charge per unit length}$$

is radially outward if the charge is positive, of magnitude

$$E = \frac{1}{4\pi\varepsilon_0} \frac{2\lambda}{r}$$

as shown in Fig. 21-3.

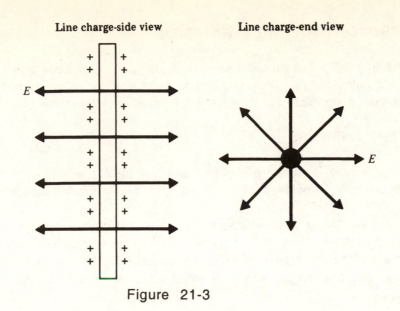

Line charge-side view   Line charge-end view

Figure 21-3

The charge density of an infinite sheet of charge, uniformly distributed, with charge density (σ) is given by:

$$\sigma = \frac{dQ}{dA} = \text{charge per unit area}$$

The electric field (E) is perpendicular to the plane, has the magnitude

$$E = \frac{\sigma}{2\varepsilon_0}$$

and is directed outward from a positive sheet, as shown in Fig. 21-4.

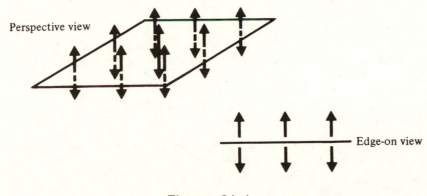

Perspective view

Edge-on view

Figure 21-4

This field has the same strength, no matter how far from the plane.

HINTS AND PROBLEM SOLVING STRATEGIES

Now is the time to review the problem-solving strategies in the main text for Coulomb's Law and Electric Field calculations. They emphasize <u>units</u> and the <u>vector</u>-nature of forces and fields, and the superposition principal for <u>summing</u> the vector contributions from charge sources.

QUESTIONS AND ANSWERS

Question. Is it possible for a charged object and a neutral object to experience an electrostatic attraction or repulsion? Consider a charged insulator and a neutral conductor.

Answer. Yes, if a positively charged insulating rod was brought near a neutral conducting shell, there would be charge separation, positive charge being repelled to the side opposite the rod and negative charge attracted to the side of the rod. This would result in a net attractive force. The same argument would hold for a negatively charged rod and an attractive force would also be obtained.

Question. Identical positive charges, Q, occupy the corners of a square of side L. What is the force on a positive test charge at the center? Explain. If the test charge was pushed slightly off-center on a diagonal, would it tend to return to the center?

Answer. At the center, the force would be zero by symmetry. You could start by assuming the test charge would be repelled more by charge 1 but since charge 1 is identical to 2,3 or 4, there is no reason for 1 to be favored. Thus the force is zero. Off center, along a diagonal, the test charge will be repelled more by the nearest charge and the net force, including that from the two charges not on that diagonal, will be back toward the center. For small displacements from the center, the test charge would behave like a simple harmonic oscillator.

EXAMPLES AND SOLUTIONS

Example 1

The force between two identical point charges 1 cm apart has a magnitude of 2 N. What is the magnitude of the point charges?

Solution:

Using Coulomb's law, one can solve for the charge in terms of the force and separation:

$$F = \frac{kq^2}{r^2} \quad \text{where} \quad k = \frac{1}{4\pi\varepsilon_0}$$

$$q = \sqrt{\frac{Fr^2}{k}} = \sqrt{\frac{(2 \text{ N})(0.01 \text{ m})^2}{8.99 \times 10^9 \text{ Nm}^2/\text{C}^2}} = \pm 1.49 \times 10^{-7} \text{ C}.$$

Since the charges are identical they are both positive or both negative. The force is repulsive in either case. It is evident that tiny fractions of a coulomb exert sizable forces at macroscopic distances. Note the importance of units. The force equation is written using a dimensional constant k, and is correct only if length is expressed in meters and charge is in coulombs.

Example 2

Electrons are removed from an originally neutral sphere and placed on another originally neutral sphere. When 1 cm apart the *small* spheres attract each other with a force of 10^{-6} N. How many electrons were transferred?

Solution:

The magnitude of the repulsive force is

$$F = \frac{k[qq']}{r^2} \quad \text{with } q = -q' = ne; \text{ where } e = 1.60 \times 10^{-19} \text{ C and } k = \frac{1}{4\pi\varepsilon_0}$$

where n is the number of electrons removed and e is the electronic charge. (The charges are equal and opposite because the surplus of electrons on one is equal to the deficit on the other.)

$$F = k\frac{[ne]^2}{r^2} \qquad n = \sqrt{\frac{r^2 F}{ke^2}}$$

Thus the number of electrons is

$$n = 6.59 \times 10^8.$$

Example 3

What is the total charge of all electrons in a gram-atom of hydrogen? If two such charges were separated by a kilometer, what would be the force between them? Convert this force to lb.

Solution:

A gram-atom or mole of hydrogen has a mass of 1.008 g and contains Avogadro's number (N_A) of atoms, with a total charge of $N_A e = 6.02 \times 10^{23} \times 1.6 \times 10^{-19}$ C $= 9.63 \times 10^4$ C. Two such charges separated by one km repel each other with a force $= 8.34 \times 10^{13}$ N, as you may find using Coulomb's law. Since a lb is $(2.2)^{-1}$ of the weight of a kg or

$$1 \text{ lb} = (2.2)^{-1}(1 \text{ kg})(9.8 \text{ m/s}^2) = 4.45 \text{ N}.$$

we find the total force in lb:

$$F = \left(8.34 \times 10^{13} \text{ N}\right) \cdot \left(\frac{1 \text{ lb}}{4.45 \text{ N}}\right) = 1.87 \times 10^{13} \text{ lb} = 9.35 \times 10^9 \text{ tons}.$$

Example 4

Positive point charges of equal magnitude $Q = 10^{-8}$ C are distributed at the points with coordinates $(0,a)$ and $(0,-a)$ in the x-y plane, with $a = 1$ cm. Find the total force these two charges exert on a negative charge $-Q$ located on the x axis at the point with coordinates $(x,0)$. Evaluate this force for $x = 0$ and $x = a$.

Solution:

First we draw a sketch, indicating the <u>vector</u> forces that the charges Q exert <u>on</u> $-Q$. Referring to Fig. 21-5,

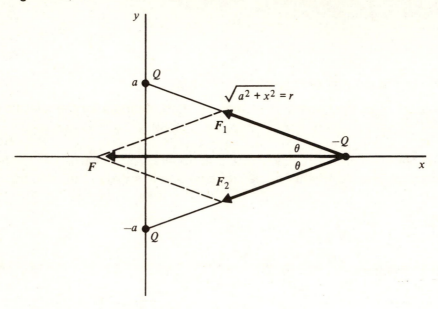

Figure 21-5

we see that each of the positive charges on the y axis attracts the negative charge on the x axis with a force

$$\left|\vec{F}_1\right| = \left|\vec{F}_2\right| = \frac{Q^2}{4\pi\varepsilon_0 r^2} = \frac{Q^2}{4\pi\varepsilon_0\left(a^2 + x^2\right)}$$

The total force on the x axis charge is the *vector* sum of the two forces shown,

$$\vec{F} = \vec{F}_1 + \vec{F}_2. \qquad F_y = 0 \quad \text{because } F_{1y} = -F_{2y}.$$

$$F_x = -2\left|\vec{F}_1\right|\cos\theta = \frac{-2Q^2}{4\pi\varepsilon_0\left(a^2 + x^2\right)}\frac{x}{\sqrt{a^2 + x^2}}$$

$$F_x = \frac{-2Q^2x}{4\pi\varepsilon_0\left(a^2 + x^2\right)^{3/2}}$$

At the origin $(x = 0)$ $F_x = 0$. At the point $(a,0) = (1\ \text{cm},0)$,

$$F_x = \frac{-2Q^2a}{4\pi\varepsilon_0\left(2a^2\right)^{3/2}} = \frac{-2Q^2}{4\pi\varepsilon_0\left(2\right)^{3/2}a^2}$$

$$F_x = \frac{-2\left(8.99 \times 10^9\ \text{Nm}^2/\ \text{C}^2\right)\left(10^{-8}\text{C}\right)^2}{\left(2\right)^{3/2}\left(0.01\ \text{m}\right)^2}$$

$$= -6.36 \times 10^{-3}\ \text{N}.$$

Note: we did the algebra for the general case then inserted numbers for the specific case.

Example 5

In the last problem find the total force of the charges at $(0, \pm a)$ on a negative charge of equal magnitude located on the y-axis at the point $(0,2a)$ and at the point $(0,a/2)$. As before $a = 1$ cm.

Solution:
The sketches below illustrate the two situations. It is important to remember that you want the forces that the Q's exert on $-Q$.

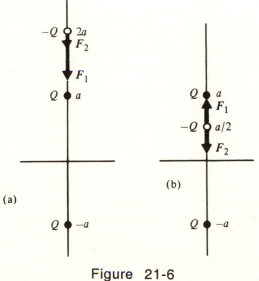

Figure 21-6

The key here is determining the correct directions of the forces. The forces on -Q at $(0,2a)$, as shown in Fig. 21-6a, both point in the negative y direction,

9

$$\vec{F} = \vec{F}_1 + \vec{F}_2. \qquad F_x = 0.$$

$$F_y = F_{1y} + F_{2y} = \frac{-Q^2}{(4\pi\varepsilon_0)a^2} + \frac{-Q^2}{(4\pi\varepsilon_0)(3a)^2}$$

$$F_y = \left(\frac{-10}{9}\right)\frac{Q^2}{(4\pi\varepsilon_0)a^2}$$

$$F_y = \left(\frac{-10}{9}\right)\frac{(8.99 \times 10^9 \ Nm^2/C^2)(10^{-8}C)^2}{(0.01 \ m)^2}$$

$$F_y = -9.99 \times 10^{-3} \ N.$$

When the charge $-Q$ is between the two Q's, the forces on $-Q$ are opposite in direction. The force on $-Q$ at $(0, a/2)$, as shown in Fig. 21-6b, has a contribution in the positive y direction from the charge Q at $(0,a)$ and in the negative y direction from the charge Q at $(0,-a)$. The total force is

$$F_y = \frac{Q^2}{(4\pi\varepsilon_0)(a/2)^2} + \frac{-Q^2}{(4\pi\varepsilon_0)(3a/2)^2} = \frac{Q^2}{(4\pi\varepsilon_0)(a)^2}\left(4 - \frac{4}{9}\right)$$

$$F_y = \left(\frac{32}{9}\right)\frac{Q^2}{(4\pi\varepsilon_0)(a)^2}$$

$$F_y = \left(\frac{32}{9}\right)\frac{(8.99 \times 10^9 \ Nm^2/C^2)(10^{-8}C)^2}{(0.01 \ m)^2}$$

$$= 3.20 \times 10^{-2} \ N.$$

Example 6

Two small pith balls of mass 5 g each are suspended by light thread of length 30 cm from a common point. They are both negatively charged and repel each other, remaining 4 cm apart at equilibrium. Find the excess number of electrons on each ball. **Note:** This problem reviews our "statics skills" and stresses the vector nature of the electrical force.

Solution:

Referring to Fig. 21-7,

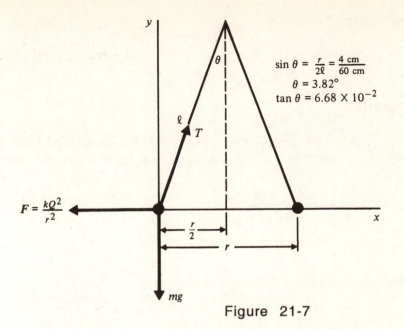

$$\sin \theta = \frac{r}{2\ell} = \frac{4 \text{ cm}}{60 \text{ cm}}$$
$$\theta = 3.82°$$
$$\tan \theta = 6.68 \times 10^{-2}$$

Figure 21-7

we see that the ball at the origin experiences forces of tension, weight, and the Coulomb repulsion of the other pith ball. The equilibrium conditions on the isolated ball at the origin are:

$$\sum F_x = 0 = T \sin \theta - \frac{1}{4\pi\varepsilon_0}\frac{Q^2}{r^2}$$

$$\sum F_y = 0 = T \cos \theta - mg .$$

Eliminating T we have

$$\tan \theta = \frac{1}{4\pi\varepsilon_0}\frac{Q^2}{mgr^2} = \frac{1}{4\pi\varepsilon_0}\frac{(ne)^2}{mgr^2}$$

where n is the number of excess electrons and Q = ne. Thus we find

$$n^2 = \left(\frac{Q}{e}\right)^2 = \frac{(4\pi\varepsilon_0)mgr^2\tan\theta}{e^2}$$

$$n^2 = \frac{(5 \times 10^{-3} \text{ kg})(9.8 \text{ m/ s}^2)(0.04 \text{ m})^2(6.68 \times 10^{-2})}{(8.99 \times 10^9 \text{ Nm}^2/ \text{ C}^2)(1.6 \times 10^{-19} \text{ C})^2}$$

$$n = 1.5 \times 10^{11} \text{ electrons.}$$

It takes lots of electrons to achieve a macroscopic effect!

Example 7

An electron rotates in a circular orbit about a heavy fixed proton. The radius of the orbit is 10^{-8} cm. Find the velocity of the electron.

Solution:

This is uniform circular motion, with the centripetal force supplied by the Coulomb attraction. Setting this force equal to the mass times the centripetal acceleration, we find

$$\frac{mv^2}{r} = \frac{1}{4\pi\varepsilon_0} \frac{e^2}{r^2}$$

$$v^2 = \frac{1}{4\pi\varepsilon_0} \frac{e^2}{m\,r}$$

$$v = \sqrt{\frac{1}{4\pi\varepsilon_0} \frac{e^2}{m\,r}}$$

$$v = \sqrt{\frac{\left(8.99 \times 10^9 \text{ N·m}^2/\text{C}^2\right)\left(1.6 \times 10^{-19} \text{ C}\right)^2}{\left(9.11 \times 10^{-31} \text{ kg}\right)\left(10^{-10} \text{ m}\right)}}$$

$$= 1.59 \times 10^6 \text{ m/s}.$$

This velocity is less than a percent of the velocity of light, so we don't have to worry about relativistic corrections.

Example 8

Find the electric field at a distance 1 cm from a positive point charge of magnitude $q = 10^{-10}$ C.

Solution:

The force on a test charge q' a distance r from q is

$$F = \frac{1}{4\pi\varepsilon_0} \frac{qq'}{r^2} = \frac{kqq'}{r^2}.$$

The electric field is the force per unit test charge, or

$$E = \frac{F}{q'} = \frac{kq}{r^2} = \frac{\left(9 \times 10^9 \text{ N·m}^2/\text{C}^2\right)\left(10^{-10} \text{ C}\right)}{\left(0.01 \text{ m}\right)^2}$$

E = 9000 N/C.

The direction of the field is the direction of the force a positive test charge at rest would experience; it is radially *outward* from q in this case.

Example 9

A positive charge of magnitude $q = 10^{-9}$ C is placed at a point where the electric field **E** has components: $E_x = 10^7$ N/C and $E_y = E_z = 0$.

     (a) Find the force on the charge.
     (b) What is the force on a negative charge of the same magnitude?

Solution:

If we know the electric field at a point, then we can always find the electric force on a known charge placed at that point:

(a) $\vec{F} = q\vec{E}$; $F_x = qE_x$; $F_y = F_z = 0$

     $F_x = (10^{-9} \text{ C})(10^7 \text{ N/C}) = 10^{-2}$ N

(b) $F_x = -10^{-2}$ N; $F_y = F_z = 0$

Example 10

Two positive point charges of magnitude $Q = 10^{-10}$ C are on the y axis at positions (0, a) and (0,-a) where a = 3 cm.
(a) Find the electric field on the x axis at the points:
(x = 1 cm, y = 0) and (x = 10 cm, y = 0).
(b) Find the electric field on the y axis at the point with coordinates (0,2 cm) and (0,6 cm).

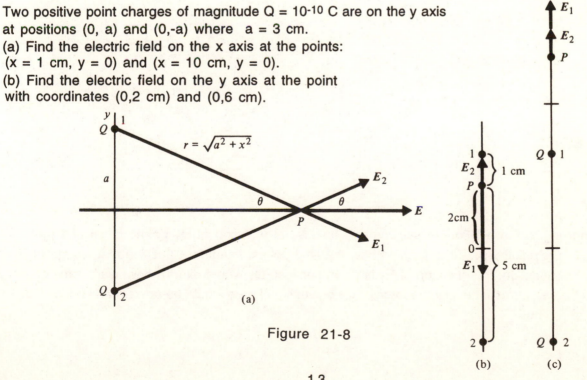

Figure 21-8

13

Solution:

Each of the two charges independently produces an electric field at the field point P.

(a) Referring to Fig. 21-8a, we calculate the field on the x axis as the *vector sum* of the field $\vec{E}_1$ due to the upper charge and the field $\vec{E}_2$ due to the lower charge.

$$\vec{E} = \vec{E}_1 + \vec{E}_2$$

$$\left|\vec{E}_1\right| = \left|\vec{E}_2\right| = \left(\frac{1}{4\pi\varepsilon_0}\right)\frac{Q}{r^2} = \left(\frac{1}{4\pi\varepsilon_0}\right)\frac{Q}{(a^2 + x^2)}$$

$$E_x = E_{1x} + E_{2x} = |E_1|\cos\theta + |E_2|\cos\theta$$

$$E_x = 2\left|\vec{E}_1\right|\cos\theta = \left(\frac{1}{4\pi\varepsilon_0}\right)\frac{2Q}{(a^2 + x^2)}\frac{x}{\sqrt{a^2 + x^2}}$$

$$E_x = \left(\frac{1}{4\pi\varepsilon_0}\right)\frac{2Qx}{(a^2 + x^2)^{3/2}} \qquad \text{where } a = 3 \text{ cm.}$$

$$E_y = E_{1y} + E_{2yi} = |E_1|\sin\theta - |E_2|\sin\theta = 0.$$

To evaluate the field here, we must recall that the distances here need to be expressed in meters. At (1 cm,0)

$$E_x = \frac{(2)(9 \times 10^9 \text{ N·m}^2/\text{C}^2)(10^{-10} \text{ C})(0.01 \text{ m})}{((0.03 \text{ m})^2 + (0.01 \text{ m})^2)^{3/2}}$$

$$= 5.69 \times 10^2 \text{ N/C}$$

At (10 cm,0)

$$E_x = \frac{(2)(9 \times 10^9 \text{ N·m}^2/\text{C}^2)(10^{-10} \text{ C})(0.1 \text{ m})}{((0.03 \text{ m})^2 + (0.1 \text{ m})^2)^{3/2}}$$

$$= 1.58 \times 10^2 \text{ N/C.}$$

(b) Referring to Fig. 21-8b, we see that the field $\vec{E}_1$ at (0,2 cm) of Q_1 points down and the field $\vec{E}_2$ of Q_2 points up. (To determine how the field points, imagine a positive test charge at P, the point in question; the direction of E_1 is down because Q_1 would exert a downward force on such a charge; correspondingly the force of Q_2 on such a charge would be upward and hence E_2 points up.) Thus

$$E_y = E_{1y} + E_{2y} = \left(\frac{1}{4\pi\varepsilon_0}\right)\frac{Q}{r_1^2} - \left(\frac{1}{4\pi\varepsilon_0}\right)\frac{Q}{r_2^2} = \frac{Q}{4\pi\varepsilon_0}\left[\frac{1}{r_1^2} - \frac{1}{r_2^2}\right]$$

$$E_y = (9 \times 10^9 \text{ Nm}^2/\text{C}^2)(10^{-10}\text{C})\left[\frac{1}{(0.01 \text{ m})^2} - \frac{1}{(0.05 \text{ m})^2}\right] = 8.64 \times 10^3 \text{ N/C}.$$

The fields at (0,6cm), on the other hand point in the same direction. Referring to Fig. 21-8c, we find for this case that

$$E_y = E_{1y} + E_{2y} = \left(\frac{1}{4\pi\varepsilon_0}\right)\frac{Q}{r_1^2} + \left(\frac{1}{4\pi\varepsilon_0}\right)\frac{Q}{r_2^2} = \frac{Q}{4\pi\varepsilon_0}\left[\frac{1}{r_1^2} + \frac{1}{r_2^2}\right]$$

$$E_y = (9 \times 10^9 \text{ Nm}^2/\text{C}^2)(10^{-10}\text{C})\left[\frac{1}{(0.03 \text{ m})^2} + \frac{1}{(0.09 \text{ m})^2}\right] = 1.11 \times 10^3 \text{ N/C}.$$

The directions of the fields caused by the charges depends on the <u>sign</u> of the charges <u>and</u> are vectors.

Example 11

If in Fig. 21-8a, the charge at position 2 (on the negative y axis) is -2Q instead of Q, find the field on the x axis as a function of x.

Solution:
We have less symmetry in this problem, so we must work harder to find the answer. A good diagram is particularly useful. Referring to Fig. 21-9

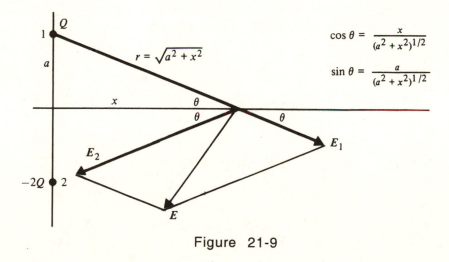

Figure 21-9

we have $\qquad \cos\theta = \dfrac{x}{\sqrt{a^2 + x^2}}$ and $\sin\theta = \dfrac{a}{\sqrt{a^2 + x^2}}$

15

Thus

$$\vec{E} = \vec{E}_1 + \vec{E}_2$$

$$E_x = E_{1x} + E_{2x} = |\vec{E}_1|\cos\theta - |\vec{E}_2|\cos\theta$$

$$E_x = \left(\frac{1}{4\pi\varepsilon_0}\right)\left(\frac{Q}{a^2 + x^2}\right)\left(\frac{x}{\sqrt{a^2 + x^2}}\right) - \left(\frac{1}{4\pi\varepsilon_0}\right)\left(\frac{2Q}{a^2 + x^2}\right)\left(\frac{x}{\sqrt{a^2 + x^2}}\right)$$

$$E_x = -\left(\frac{1}{4\pi\varepsilon_0}\right)\frac{Qx}{\left(a^2 + x^2\right)^{3/2}}$$

$$E_y = E_{1y} + E_{2y} = -|E_1|\sin\theta - |E_2|\sin\theta$$

$$E_y = \left(\frac{1}{4\pi\varepsilon_0}\right)\left(\frac{-Q}{a^2 + x^2}\right)\left(\frac{a}{\sqrt{a^2 + x^2}}\right) + \left(\frac{1}{4\pi\varepsilon_0}\right)\left(\frac{-2Q}{a^2 + x^2}\right)\left(\frac{a}{\sqrt{a^2 + x^2}}\right)$$

$$E_y = \left(\frac{1}{4\pi\varepsilon_0}\right)\frac{-3Qa}{\left(a^2 + x^2\right)^{3/2}}$$

$$E = \sqrt{E_x^2 + E_y^2} = \frac{kQ}{\left(a^2 + x^2\right)^{3/2}}\sqrt{9a^2 + x^2}$$

There are no numbers to plug in here! It sure was important to keep signs of the various components straight.

Example 12

An electron with initial velocity v_0 (in the +x direction) enters a region of space between two conducting plates, 2 cm long and 1 cm wide, which has a uniform electric field (**E**) of magnitude 5000 N/C in the +y direction as shown in Fig. 21-10. What is the minimum velocity (v_0) it must have to avoid hitting one of the plates?

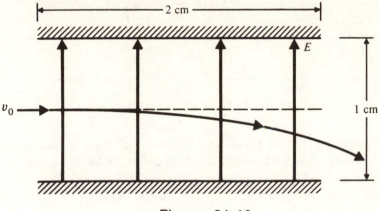

Figure 21-10

Solution:

We take the y axis to be vertical and the x axis to be horizontal. If the electron moves too slowly, it will hit the bottom plate. The electric force on the electron is

$$\vec{F} = q\vec{E} = -e\vec{E}$$

and thus is directed down in Fig. 21-10, opposite to the field direction. With the origin of the x-y system at the center of the left opening, the equations of motion for the electron are

$$F_x = 0 = ma_x \qquad a_x = 0 \qquad x = x_0 + v_{0x}t$$

$$F_y = -eE = ma_y \qquad\qquad a_y = -eE/m$$

$$y = y_0 + v_{0y}t - \frac{eE}{2m}t^2$$

Since $x_0 = 0$, $y_0 = 0$, $v_{0x} = v_0$, and $v_{0y} = 0$, and $x = v_0t$, we have

$$y = -\frac{eE}{2m}t^2 = \frac{-eE}{2m}\left(\frac{x}{v_0}\right)^2 .$$

To just hit the plate as shown, x = 2 cm when y = -0.5 cm. Solving the last expression for v_0,

$$v_0 = \sqrt{\left(\frac{-eE}{2m}\right)\left(\frac{x^2}{y}\right)}$$

17

where y is negative, so the expression under the square root is positive. Using the values

$$e = 1.6 \times 10^{-19} \text{ C}; \quad m = 9.1 \times 10^{-31} \text{ kg}; \quad E = 5000 \text{ N/C}.$$

We find

$$v_0 = \sqrt{\left(\frac{-\left(1.6 \times 10^{-19} \text{ C}\right)\left(5000 \text{ N/C}\right)}{2\left(9.1 \times 10^{-31} \text{ kg}\right)}\right)\left(\frac{(0.02 \text{ m})^2}{(-0.005 \text{ m})}\right)}$$

$$= 5.93 \times 10^6 \text{ m/s}.$$

Example 13

A uniform electric field of magnitude E exists between two conducting plates separated by a distance d. An electron is released from rest at the negative plate and is attracted to the positive plate.

(a) Find its kinetic energy when it collides with the positive plate, in terms of E, d, and the electronic charge.

(b) If E = 10 N/C and d = 1 cm, find the velocity of impact.

Solution:

Since the force is constant, so is the acceleration, and we can use our previous kinematical expressions for constant acceleration Referring to Fig. 21-11, we have

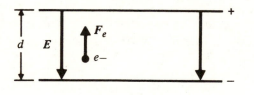

Figure 21-11

$$F_e = eE = ma \quad \text{or} \quad a = eE/m, \text{ using Newton's Second Law.}$$

For constant acceleration

$$v^2 = v_0^2 + 2ax.$$

If the electron starts from rest ($v_0 = 0$) and travels a distance d,

$$v^2 = 2ad = 2(eE/m)d$$

Its final kinetic energy is

$$K = \frac{1}{2} mv^2 = eEd$$

(b)   Its velocity is

$$v = \sqrt{\frac{2eEd}{m}}$$

$$v = \sqrt{\left(\frac{2\left(1.6 \times 10^{-19} \text{ C}\right)\left(10 \text{ N/C}\right)\left(0.01 \text{ m}\right)}{\left(9.1 \times 10^{-31} \text{ kg}\right)}\right)}$$

$$= 1.88 \times 10^5 \text{ m/s}.$$

Example 14

Find the electric field on the axis of and a distance z away from a circular disk of radius a and of uniform charge density σ per unit area.

Solution:

This is a good problem for sharpening vector calculus skills. Occasionally, graduate students will stumble on this one, so let's take it slowly. Referring to Fig. 21-12, we divide the disk into thin concentric rings of thickness dr, as shown.

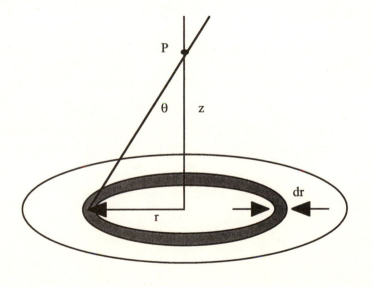

Figure 21-12

Each ring has an area dA = 2πr dr and hence a total charge

19

$dQ = \sigma \, dA = \sigma 2\pi r \, dr.$

Such a ring of charge contributes an axial field at z of magnitude

$$dE_z = \left(\frac{1}{4\pi\varepsilon_0}\right)\left[\frac{\sigma \, 2\pi r \, dr}{z^2 + r^2}\right]\cos\theta \quad \text{where} \quad \cos\theta = \frac{z}{\sqrt{z^2 + r^2}}$$

To find the total field, sum all ring contributions,

$$E_z = \frac{1}{4\pi\varepsilon_0}(2\pi z\sigma)\int_0^a \frac{r \, dr}{\left(z^2 + r^2\right)^{3/2}}$$

With the substitution

$$x = r^2 \qquad dx = 2r dr$$

the integral is

$$E_z = \frac{z\sigma}{2\varepsilon_0}\int_0^{a^2} \frac{(1/2) \, dx}{\left(z^2 + x\right)^{3/2}} = \frac{z\sigma}{4\varepsilon_0}\left.\frac{\left(z^2 + x\right)^{-1/2}}{(-1/2)}\right|_0^{a^2}$$

Note for $z \ll a$ the field approaches that of an infinite sheet,

$$\text{Lim}_{z\to 0} \, E_z = \frac{\sigma}{2\varepsilon_0} = \text{field of infinite sheet.}$$

Example 15

A thin wire of length L carries λ charge per unit length. Find the electric field on the axis of the wire a distance a from an end. What is the magnitude of this field when $L/a \ll 1$?

Solution:

Referring to Fig. 21-13,

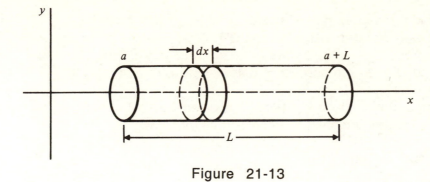

Figure 21-13

we first find the field at the origin due to the bit of thin wire at dx, carrying a charge $\lambda\,dx$:

$$dE_x = \frac{1}{4\pi\varepsilon_0}\frac{\lambda\,dx}{x^2}$$

The total field is

$$E_x = \frac{-\lambda}{4\pi\varepsilon_0}\int_a^{L+a}\frac{dx}{x^2}$$

$$E_x = \frac{-\lambda}{4\pi\varepsilon_0}\left.\frac{-1}{x}\right|_a^{L+a} = \frac{-\lambda}{4\pi\varepsilon_0}\left[\frac{1}{a} - \frac{1}{L+a}\right]$$

For $L/a \ll 1$, the field has the magnitude

$$E_x = \frac{-\lambda}{4\pi\varepsilon_0}\left[\frac{1}{a} - \frac{1}{a(1+L/a)}\right]$$

$$E_x = \frac{-\lambda}{4\pi\varepsilon_0}\frac{1}{a}\left[1 - \left(1+\frac{L}{a}\right)^{-1}\right] \cong \frac{-\lambda}{4\pi\varepsilon_0}\frac{1}{a}\left[1 - 1 + \frac{L}{a} + ...\right]$$

$$E_x \cong \frac{-\lambda L}{4\pi\varepsilon_0 a^2} = \frac{-Q}{4\pi\varepsilon_0 a^2}$$

where $Q = \lambda L$ is the net charge on the wire. The minus sign indicates that the field points in the negative x direction. Far away, the piece of wire looks like a point charge.

QUIZ

1. At a point 1.3 of the way along a line from charge Q to charge Q', a third charge experiences no electrical force.
　　　(a) What is the relative sign of Q and Q'?
　　　(b) What is the relative size of Q and Q'?

Answer: (a) Q and Q' are both positive or both negative. (b) Q/Q' = 1/4

2. A positive charge of magnitude 2×10^{-9} C is located 4×10^{-2} m from a negative charge of magnitude 3×10^{-9} C. What is the force between the charges?

Answer: 0.34×10^{-9} N, attractive

3. Calculate the ratio of the Coulomb force to the gravitational force of two electrons for each other.

Answer: 4.2×10^{42}

4. Two opposite charges of magnitude 2×10^{-14} C are separated by 1 cm. Find the electric field on the line passing through the charges, 1 m from the midpoint of the charges.

Answer: 3.6×10^{-6} N/C. The field is along the line passing through the points, in the negative to positive direction.

5. Find the electric field on the perpendicular bisector of a uniformly distributed line of charge of magnitude Q and length a as a function of the distance x from the charge.

Answer: $\{Q/4\pi\varepsilon_0 x[(a/2)^2 + x^2]^{1/2}\}$

22
GAUSS'S LAW

OBJECTIVES

In this short chapter, based on the Coulomb force law, electric flux is introduced and the field strength at a point is calculated by use of Gauss's law in charge distributions of high symmetry. Your objectives are to:

Calculate the *electric flux* through a surface.

By use of *Gauss's Law*, calculate the electric field of symmetrical charge distributions, such as charged surfaces of spherical and cylindrical *conductors*, and of continuous volume charge distributions of insulating bodies with high symmetry.

Establish that stationary charge may exist only on the *surfaces* of conductors. Establish that the static *electric field always is zero within a conductor*, and perpendicular to the conducting surface, outside of the conductor.

REVIEW

Electric flux Φ_E is the product of the perpendicular component of the electric field $E_\perp$ through a surface of area A. When the field is uniform over the surface, $\Phi_E = E_\perp A$. When the field varies over the surface, the flux is calculated from a surface integral. Flux takes its sign from the direction of the electric field and is counted positive when it comes out of a closed surface. Gauss's law is a relation, following from Coulomb's law, for the net flux through a closed surface:

$$\int E_n \, dA = \oint \vec{E} \cdot d\vec{A} = \frac{Q_{enc}}{\varepsilon_0}$$

Once mastered, Gauss's law provides a simple and powerful method of calculating the electric fields of symmetrical charge distributions without the labor of direct integration.

The quantity $E_n dA = \vec{E} \cdot d\vec{A}$ is the product of an area increment and the perpendicular component of field pointing through it. $E_n dA$ is counted positive if the field points out of the enclosed surface in the integral and negative otherwise.

The symbol $\oint$ means that the integral is to be done over a *closed surface* and Q_{enc} is the total electric charge inside that closed surface. In applying this law, we frequently make use of the facts: $E_n dA = 0$ when the field is parallel to the surface and $E_n dA = EdA$ when the field is perpendicular to the surface.

Care must be taken in distinguishing between charges on a conductor (which always reside on a surface of the conductor in the form of σ = charge per unit area) and idealized charge distributions in insulating bodies which may take the forms:

$$\rho = \frac{dQ}{dV} = \text{charge per unit volume (volume charge)}$$

$$\sigma = \frac{dQ}{dA} = \text{charge per unit area (surface or layer charge)}$$

$$\lambda = \frac{dQ}{dL} = \text{charge per unit length (line charge)}$$

$$Q = \text{point charge.}$$

Using Gauss's law it can be easily shown that the electric field near a conducting surface is given by $E = \sigma/\varepsilon_0$ and is always perpendicular to the surface.

HINTS AND PROBLEM-SOLVING STRATEGIES

Now you should review the Gauss's Law Problem-Solving Strategies of the main text, which we will apply in the examples below.

QUESTIONS AND ANSWERS

Question. Gauss's law is always true but it not always useful. When can you benefit from using it?

Answer. To benefit from Gauss's law, you must be able to find a surface through the point of interest, that you can describe mathematically, upon which the electric field is parallel to the normal and of the same magnitude all over the surface. The, the scalar product $\vec{E} \cdot \hat{n}$ can be removed from the surface integral leaving only the surface area and the enclosed charge to be evaluated before E can be calculated.

Question. Can you verify the following statements using lines of force or Gauss's law? (a) the field of a uniformly charged sphere is $4\pi\varepsilon_0 E = Q/r^2$ outside the sphere; (b) the field inside a uniformly charged sphere or cylindrical shell is zero; (c) the field everywhere inside a conductor is zero;

Answer. (a) Use Gauss's Law. Draw a sphere of radius r > R where R is the charged sphere radius. The flux through the construction sphere surface is Q/ε_0 and the surface area is $4\pi r^2$. (b) Use Gauss's Law or lines of force. With lines of force, it is obvious that any field that exists must be along the radius. Draw an arrow, representing the field direction, along a radius. Now look at the radius to the opposite side. Draw a corresponding arrow along that radius. These arrows point in opposite directions at the center so the field would not be single valued. (c) Use

Gauss's Law. Since the excess charges can only be on the outside surface of a conductor the flux through <u>any</u> closed surface inside a conductor is zero making E = 0.

From the center of the sphere, the lines of E could point radially outward from C or radially inward toward C. Since lines of force <u>cannot cross</u>, in either case, there is a problem at point C since the two lines <u>do</u> cross there. Thus since our assumption that lines of force exist inside the sphere leads to a contradiction, there can be no field inside the sphere.

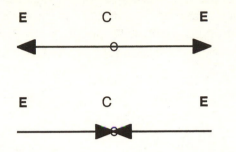

Using Gauss's law, since there is no enclosed charge for any sphere of radius smaller than the given sphere's radius, there is no flux through the gaussian surface and thus no electric field.

EXAMPLES AND SOLUTIONS

Example 1

Show that the electric field <u>near a conductor</u> is perpendicular to the conductor and has magnitude $E = \sigma/\varepsilon_0$.

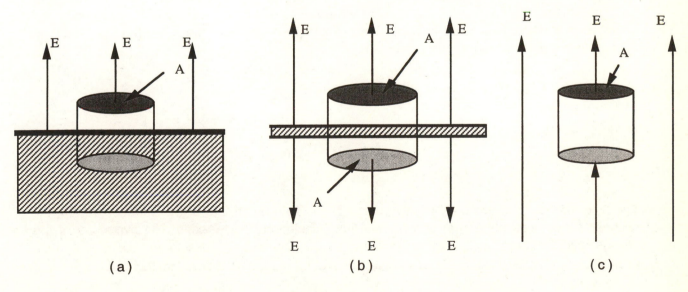

(a)   (b)   (c)

Figure 22-1

25

Solution:

First we make a physical observation: the field must be perpendicular; if it had a tangential component, free charges in the conductor would separate and move until a static situation is reached where all fields inside the conductor are zero.

Then we choose a Gaussian Surface that exploits this field direction. Consider now a closed surface in the shape of a right cylinder whose axis is along the field lines (see Fig. 22-1a) with top of the cylinder (area A) just outside the conductor surface and the bottom just inside. The total flux out of the cylinder has contribution only from the top cap with flux $\Phi_E = E_\perp A = EA$. (The side does not contribute because E has no perpendicular component there.) This must be equal to the net charge enclosed in the cylinder divided by ε_0:

$$\Phi_E = E_\perp A = EA = \sigma A/\varepsilon_0$$

or

$$E = \frac{\sigma}{\varepsilon_0}.$$

Example 2

Use Gauss's law to find the field of an infinite sheet of charge with density σ = charge per unit area.

Solution:

Refer to Fig. 22-1b. Think of the charge as embedded in a thin insulator; there are no conductors in sight. First observe from the symmetry of the problem, the field direction must be perpendicular to the sheet. The layer of charge produces a field as shown on both sides of the sheet so that the outward flux is twice that of Example 1 because there are flux contributions from both bottom and top caps of the cylinder,

$$\Phi_E = 2\,EA = \frac{\sigma A}{\varepsilon_0}$$

which gives for E

$$E = \frac{\sigma}{2\varepsilon_0}.$$

The flux contributions are both positive because the field points out of the surface. Again, the cylinder shown was chosen to simplify the flux calculation and to exploit the symmetry of the charge distribution.

Example 3

Use Gauss's law to show that there may be no charge anywhere in a region where $\vec{E}$ is uniform.

Solution:

Again use the right cylindrical mathematical surface as in Examples 1 and 2. (See Fig. 22-1c.) At any point inside the region of uniform field

$$\Phi_E = (EA)_{top} + (EA)_{bottom}$$

because the sides of the cylinder make no contribution as in the previous examples. Now the lines of E enter the bottom surface (negative flux) and exit the top (positive flux) and the two terms cancel,

$$\Phi_E = 0 = \frac{Q_{enc}}{\varepsilon_0}$$

so that the charge enclosed is zero at any point inside the region where E is uniform.

Example 4

A long cylindrical solid conductor has radius R and λ charge per unit length. Find the field inside and outside of the conductor.

Solution:

Inside the conductor the field is zero. Charges on conductors reside only on the surface.

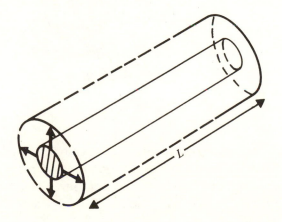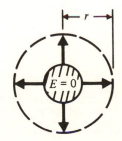

Figure 22-2

27

Referring to Fig. 22-2, we apply Gauss's law to a cylindrical surface coaxial with the charge conductor of radius r > R. By the cylindrical symmetry the field must point radially outward. Thus

$$\oint E_\perp \, dA = EA = E(2\pi rL) = \frac{Q}{\varepsilon_0} = \frac{\lambda L}{\varepsilon_0}$$

$$E = \frac{2\lambda}{4\pi r\varepsilon_0} \qquad r > R$$

$$E = 0 \qquad r < R$$

The exterior field is same as that of a *thin* wire on the cylinder axis, with the same charge per unit length. Note again that the choice of the Gaussian Surface was dictated by the symmetry of the charge distribution.

Example 5

A long *insulating* cylinder of radius R contains a uniform distribution of charge density per unit volume ρ. Find the electric field inside and outside of the cylinder.

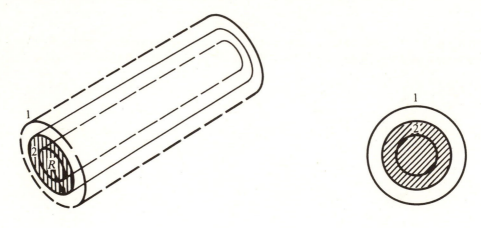

Figure 22-3

Solution:

Now there may be charge inside the Gaussian Surface because we are no longer dealing with a conductor. Referring to Fig. 22-3 we apply Gauss's law in the two regions inside and outside of the (crosshatched) uniformly charged cylinder. As before the field is radial and constant on a surface of fixed radius r, and end caps have no contribution because $E_n dA = 0$ on them.

For r > R we have, applying Gauss's Law to surface 1, outside of the charged cylinder

$$\oint E_\perp \, dA = EA = E(2\pi rL) = \frac{Q}{\varepsilon_0}$$

The charge enclosed Q is the charge density ρ times the volume containing the charge

$$Q = \pi \rho R^2 L$$

and thus

$$E(2\pi rL) = \frac{\rho \pi R^2 L}{\varepsilon_0} \qquad E = \frac{\rho R^2}{2r\varepsilon_0}.$$

For $r < R$ we have, applying Gauss's Law to surface 2,

$$\oint E_\perp \, dA = EA = E(2\pi rL).$$

However, the charge enclosed Q' is now less than before,

$$Q' = \rho V' = \pi \rho r^2 L$$

since a part of the charge distribution is outside of the surface. Thus we find

$$E(2\pi rL) = \frac{Q}{\varepsilon_0} = \frac{\rho \pi r^2 L}{\varepsilon_0}$$

$$E = \frac{\rho r}{\varepsilon_0}.$$

Note at $r = R$ the inner field matches the outer field. The field is plotted in Fig. 22-4.

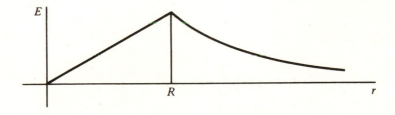

Figure 22-4

Example 6

A *conducting sphere* of radius R contains a net charge Q. Find the field inside and outside of the sphere.

Solution:

Referring to Fig. 22-5

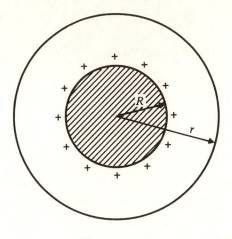

Figure 22-5

we apply Gauss's law to a spherical surface concentric with the sphere. Because we have a conductor, all of the charge resides *on the surface* of the sphere and the inner field is zero; if there were an inner field it would be radial, but then

$$\oint E_\perp \, dA = EA = \frac{Q}{\varepsilon_0} = 0$$

$$E = 0 \qquad r < R$$

verifying that the field must be zero inside the conductor. For r > R the field is radial and

$$\oint E_\perp \, dA = EA = E(4\pi r^2) = \frac{Q}{\varepsilon_0}$$

$$E = \frac{Q}{4\pi r^2 \varepsilon_0} \qquad r > R$$

Thus a *sphere* carrying a charge Q has the same field as a *point charge* Q when r > R (outside of the sphere).

Example 7

A conducting spherical shell of outer radius 2R is electrically neutral. The conductor contains a concentric spherical cavity of radius R. At the center of the cavity is suspended a point charge Q. Find the electric field for regions $0 < r < R$, $R < r < 2R$, and $2R < r$.

Solution:

Referring to Fig. 22-6, where the conductor is crosshatched,

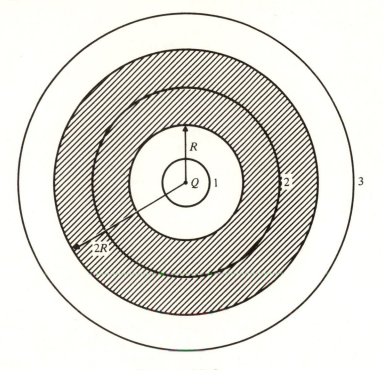

Figure 22-6

we apply Gauss's law to three concentric spherical surfaces, first to the one inside the cavity, surface (1) where $r < R$,

$$\oint_1 E_\perp \, dA = \frac{Q}{\varepsilon_0}$$

$$E(4\pi r^2) = \frac{Q}{\varepsilon_0} \qquad r < R$$

$$E = \frac{Q}{4\pi r^2 \varepsilon_0} \qquad r < R$$

Inside the conductor $E = 0$ and Gauss' law applied to surface 2, where $R < r < 2R$, implies

$$\oint_2 E_\perp \, dA = \frac{Q_{enc}}{\varepsilon_0} = 0 \qquad R < r < 2R \qquad E = 0$$

Surface 2 encloses the point charge Q. Evidently an equal but opposite charge -Q is spread over the *inner* surface of the conductor at r = R.

Since the conductor is electrically neutral, a charge Q must reside on its outer surface at r = 2R. Thus for the region outside of the conductor

$$\oint_3 E_\perp \, dA = \frac{Q_{total}}{\varepsilon_0} = \frac{Q - Q + Q}{\varepsilon_0} = \frac{Q}{\varepsilon_0}$$

$$E(4\pi r^2) = \frac{Q}{\varepsilon_0}$$

$$E = \frac{Q}{4\pi r^2 \varepsilon_0} \qquad r > 2R$$

Example 8

An insulating sphere has a uniform charge density ρ per unit volume. Find the electric field inside and outside of the sphere.

Solution:

Referring to Fig. 22-7,

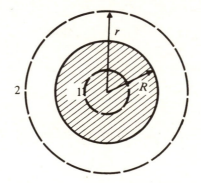

Figure 22-7

we apply Gauss's law to concentric spherical surfaces with radii r < R and r > R. The field is radial and everywhere perpendicular to these surfaces. For r < R

$$\oint_1 E_\perp \, dA = \frac{Q}{\varepsilon_0} \qquad Q = \rho V = \rho(\frac{4}{3}\pi r^3)$$

$$E\left(4\pi r^2\right) = \frac{4\rho\pi r^3}{3\varepsilon_0}$$

$$E = \frac{\rho r}{3\varepsilon_0} . \qquad r < R$$

Note in calculating Q, only the charge *within* the surface is included.

For r > R

$$\oint_2 E_\perp \, dA = \frac{Q}{\varepsilon_0} \qquad Q = \rho V = \rho\left(\frac{4}{3}\pi R^3\right)$$

$$E\left(4\pi r^2\right) = \frac{4\rho\pi R^3}{3\varepsilon_0}$$

$$E = \frac{\rho R^3}{3r^2\varepsilon_0} \qquad r > R$$

Note at r = R the expressions for the inner and outer fields match.

QUIZ

1. A long solid cylinder of charge has a nonuniform charge density which varies with the radial coordinate, $\rho = \rho_0(r/R)$ where R is the radius of the cylinder. Using Gauss's Law find the field inside the cylinder.

Answer: $E = (\rho_0 r^2/3R\varepsilon_0)$, radially outward if $\rho_0 > 0$.

2. A modest sized electric field near a conducting surface is 1250 N/C. What is the surface charge density in terms of electron per square meter?

Answer: 7×10^{10} electrons per m^2

3. A spherical shell of uniform charge density ρ has inner radius r_1 and outer radius r_2. Find the electric field in the three regions $r < r_1$, $r_1 < r < r_2$, and $r_2 < r$.

Answer: $0, \dfrac{\rho}{3\varepsilon_0 r^2}\left(r^3 - r_1{}^3\right), \dfrac{\rho}{3\varepsilon_0 r^2}\left(r_2{}^3 - r_1{}^3\right)$ and radial in direction.

23
ELECTRICAL POTENTIAL

OBJECTIVES

This chapter is based on the conservative nature of the the coulomb force. According to the analysis of Chapter 7, a *potential energy* function U may then be introduced when the forces are conservative, and the total energy K + U is conserved if no other forces act. The electrical *potential* V is the potential energy per unit charge, V = U/q. Your objectives are to:

Find the *potential* of a distribution of point charges.

Calculate the potential at a point by calculating the work per unit charge, the *line integral* of the electric field.

Evaluate the potential difference between charged *parallel plates*.

Evaluate the potential difference between charged *spherical concentric conductors*.

Evaluate the potential difference between charged *coaxial cylindrical conductors*.

Calculate the *work* done when a charge is moved in an electric field from one potential to another.

Apply the *conservation* of kinetic and *electric potential energy* to problems involving charged bodies moving in regions where there is an electric field.

REVIEW

The work done by a conservative force F on a body that moves along a path from $a \rightarrow b$ is given by the line integral

$$W_{a \rightarrow b} = \int_a^b \vec{F} \cdot d\vec{l} = \int_a^b (F \cos \theta) \, dl = U_a - U_b$$

The work of the conservative force is path independent and equal to the decrease of the potential energy:

$U_a - U_b$.

For a point charge q in the field of a point charge Q, moved from r_a to r_b, the corresponding change in potential energy is

$$W_{a \to b} = \int_a^b \vec{F} \cdot d\vec{l} = \int_a^b \frac{Qq}{4\pi\varepsilon_0} \frac{dr}{r^2} = \frac{Qq}{4\pi\varepsilon_0} \left[-\frac{1}{r} \right]_{r_a}^{r_b}$$

$$= \frac{Qq}{4\pi\varepsilon_0} \left[\left(\frac{1}{r_a} \right) - \left(\frac{1}{r_b} \right) \right] = U_a - U_b$$

The potential energy of q in the field of Q is thus

$$U = \frac{Qq}{4\pi\varepsilon_0} \left(\frac{1}{r} \right)$$

where the potential energy is referred to a base level $U = 0$ at $r = \infty$.

The *electric potential* V is the potential energy per unit charge, defined as the *difference*

$$V_a - V_b = \frac{U_a - U_b}{q} = \int_a^b \frac{\vec{F} \cdot d\vec{l}}{q} = \int_a^b \vec{E} \cdot d\vec{l}$$

The electric potential is a *scalar*, unlike the field $\vec{E}$, which is a vector.

For a point charge the potential difference between points distance r_a and r_b from the charge is:

$$V_a - V_b = \frac{Q}{4\pi\varepsilon_0} \left[\left(\frac{1}{r_a} \right) - \left(\frac{1}{r_b} \right) \right]$$

Taking $r_b = \infty$ to be a reference point where $V_b = 0$, we find the potential at any point $r = r_a$ is

$$V(r) = \frac{Q}{4\pi\varepsilon_0} \left(\frac{1}{r} \right); \quad V = 0 \text{ at } r = \infty.$$

For a *distribution of point charges* q_i at distances r_i, the potential is the sum of the point charge potentials

$$V = \sum_i \frac{q_i}{4\pi\varepsilon_0} \left(\frac{1}{r_i} \right).$$

Electric potential V and potential energy U are strictly defined only up to an additive constant. The potential V_b may be assigned any value. Often it is convenient to let b be the point at infinity and set $V_b = 0$; then

$$V_a = \int_a^\infty \vec{E} \cdot d\vec{l}$$

If the potential depends on a single radial coordinate, and the field only has a component in this radial direction, as is the case for a point charge or distributions with spherical or cylindrical symmetry, then

$$V(r) = \int_r^\infty \vec{E} \cdot d\vec{l} = \int_r^\infty E\, dr$$

For a point charge of magnitude q

$$V(r) = \int_r^\infty \frac{q\, dr}{4\pi\varepsilon_0 r^2} = \frac{q}{4\pi\varepsilon_0}\left[-\frac{1}{r}\right]_r^\infty = \frac{q}{4\pi\varepsilon_0 r}$$

The differential of the potential in general can be written in terms of the *gradient* operator:

$$dV = \mathbf{grad}\ V \cdot dl = -\vec{E} \cdot d\vec{l},$$

We retrieve the original definition of potential by integrating dV along a path,

$$\int_a^b dV = V_b - V_a = -\int_a^b \vec{E} \cdot d\vec{l}$$

$$V_a - V_b = \int_a^b \vec{E} \cdot d\vec{l}$$

Thus we have

$$\vec{E} = -\mathbf{grad}\ V$$

$$E_x = -\left(\frac{\partial V}{\partial x}\right), \qquad E_y = -\left(\frac{\partial V}{\partial y}\right) \qquad E_z = -\left(\frac{\partial V}{\partial z}\right)$$

The SI unit of electric potential is the volt (V). Useful conversions are

$$1\ V = (1\ N/C)\cdot m = 1\ J/C$$

Potential and potential differences are sometimes colloquially referred to as voltage or voltage differences.

HINTS AND PROBLEM-SOLVING STRATEGIES

It's time again to review the problem-solving strategy for calculating Electric Potential in the main text, the use of which will be illustrated in the examples below.

36

QUESTIONS AND ANSWERS

Question. An electron is 5.3 x 10⁻⁹ cm away from a proton. How would you calculate the electron's "escape" velocity?

Answer. To find the escape velocity, calculate the potential energy at the electron's position. Taking the potential energy at infinite separation to be zero, the change in potential energy is then the magnitude of the original potential energy. If the electron is given a velocity v, away from the proton, the resulting kinetic energy is equated to the change in potential energy.

Question. Is it possible to take four charges, all of the same magnitude but not necessarily the same sign, and place them at the corners of a square so that both the electric field and the electric potential (assume the zero for the point charge potential is at infinity) are zero at the square's center?

Answer. Yes. For instance, place two positive charges at the ends of one diagonal and two negative ones at the ends of the other diagonal.

EXAMPLES AND SOLUTIONS

Example 1

Two large parallel plates are separated by 0.02 m and have an electric field of magnitude 1000 N/C between them.
  (a) Find the potential difference between the plates.
  (b) Assigning the positive plate zero potential, what is the potential of the negative plate?
  (c) Can the potential at the negatively charged plate be positive?
  (d) Find the potential at a distance r from the positive plate.

Solution:

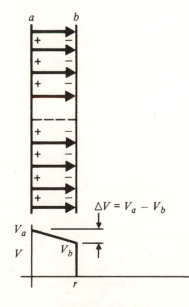

Figure 23-1

37

First, draw a sketch indicating the charges and the field directions we expect. The potential difference is the potential energy per unit (positive) charge or the work per unit charge and we take this point of view in (a).

(a) The potential (see Fig. 23-1) is given by

$$V_a - V_b = \int_a^b \vec{E} \cdot d\vec{r}$$

where for the path we choose the one for which the evaluation of the integral is simplest, namely the dashed line perpendicular to the plates and parallel to the field:

$$V_a - V_b = \int_a^b E \cdot dr = E \int_a^b dr = ER$$

where R is the plate separation. The potential difference is

$$V_a - V_b = (1000 \text{ N/C})(0.02 \text{ m}) = 20 \text{ V}.$$

Note plate a is at a higher potential because a positive charge there would be repelled or "roll down" the slope in the diagram at the bottom of Figure 24-1.

(b) If $V_a = 0$, then plate b is at potential

$$V_b = -20 \text{ V}$$

(c) If $V_b = 0$, $V_a = 20$ V.

If $V_a = 10$ V, $V_b = V_a - 20$ V $= 10$ V $- 20$ V $= -10$ V.

If $V_b = +10$ V, $V_a = 20$ V $+ V_b = 30$ V.

In the last case V_b is positive. All are possible assignments. In all cases $V_a - V_b = 20$ V. It is up to you where you want to set $V = 0$. Only ΔV is determined by the physics.

(d)

$$V_a - V_r = \int_a^r \vec{E} \cdot d\vec{r} = Er$$

$$V_r = V_a - rE = V_a - r\left(\frac{V_a - V_b}{R}\right)$$

The potential falls linearly from V_a with a slope -E, that is

$$\frac{dV_r}{dr} = -E = -\left(\frac{V_a - V_b}{R}\right)$$

V(r) is plotted in Fig. 23-1, below the parallel plates.

Example 2

Find the electrical field between two large parallel plates, 1 cm apart, with a potential difference of 220 V.

Solution:

The electric field is uniform, as in Fig. 23-1. The potential difference is

$$220 \text{ V} = V_a - V_b = \int_a^b \vec{E} \cdot d\vec{L} = EL$$

$$E = \frac{220 \text{ V}}{.01 \text{ m}} = 22{,}000 \text{ V/m} = 22{,}000 \text{ N/C}$$

Example 3

Find the potential inside and outside of a solid spherical conductor of radius R = 10 cm carrying a charge Q = 10^{-9} C. Let V = 0 at infinity. Evaluate your result for r = 5 cm and r = 20 cm. Sketch the potential as a function of r.

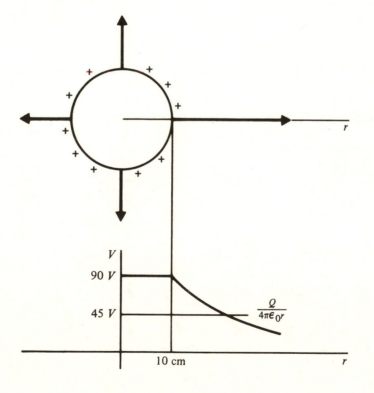

Figure 23-2

39

Solution:

Generally, to find electric potentials, we first need to find electric fields. Referring to Fig. 23-2, we note that the field is radially outward for r > R (the symmetry requires this) and zero within the conductor, r < R (the field is always zero within a conducting region.) Choosing the a radial path for the line integral,

$$V_a - V_b = \int_a^b \vec{E} \cdot d\vec{L}$$

we set $V_b = 0$ at $b = \infty$ and allow the point a to be at a general r, $V_a = V(r)$. Thus we have

$$V(r) = \int_r^\infty E \cdot dr = \int_r^\infty E_r \, dr,$$

We know the fields from our previous work: (field vanishes inside a conductor)

$$E = 0 \qquad r < R$$

$$E_r = \frac{Q}{4\pi\varepsilon_0 r^2} \qquad r > R$$

Thus for r > R the integral can be evaluated as

$$V(r) = \int_r^\infty \frac{Q}{4\pi\varepsilon_0 r^2} \, dr = \frac{Q}{4\pi\varepsilon_0}\left[-\frac{1}{r} \right]_r^\infty = \frac{Q}{4\pi\varepsilon_0 r}$$

At r = 20 cm, we find

$$V(20 \text{ cm}) = V(0.2 \text{ m}) = \frac{\left(10^{-9} \text{ C}\right)}{(4\pi)\left(8.85 \times 10^{-12} \text{ C}^2/\text{N·m}^2\right)(0.2 \text{ m})}$$

$$V = 45 \text{ V}$$

For r < R we break the integral into two parts

$$V(r) = \int_r^R \vec{E} \cdot d\vec{L} + \int_R^\infty \vec{E} \cdot d\vec{L}$$

$$V(r) = 0 + \int_R^\infty \frac{Q}{4\pi\varepsilon_0 r^2} \, dr$$

The first term is zero because E = 0 when r < R. Thus anywhere inside the solid conductor we have the same potential as at the surface,

$$V = \frac{Q}{4\pi\varepsilon_0}\left[-\frac{1}{r}\right]_R^\infty = \frac{Q}{4\pi\varepsilon_0 R}$$

$$V = \frac{\left(10^{-9}\ C\right)}{(4\pi)\left(8.85 \times 10^{-12}\ C^2/N\cdot m^2\right)\left(0.1\ m\right)}$$

$$= 90\ V.$$

A sketch of V(r) is given in Fig. 23-2b. Conductors have the same potential everywhere so long as charges don't move.

Example 4

A positive charge q of mass m is released from rest at the positive capacitor plate of Fig. 23-1.
      (a) Find the acceleration of the charge, and its velocity and kinetic energy when it strikes the other plate. Express your results in terms of V_a, V_b, q, the plate separation L, and the mass m.
      (b) Find the work done by the electric field on the charge.
      (c) Find the change in potential V and the change in potential energy U of the charge.
      (d) What are the relations among the results (a) - (c)?

Solution:

(a) The constant acceleration is

$$a = \frac{F}{m} = \frac{qE}{m} = \frac{q}{m}\left(\frac{V_a - V_b}{L}\right).$$

If v is the final velocity, and the initial velocity is equal to zero,

$$v^2 = 2aL = 2\frac{q}{m}\left(\frac{V_a - V_b}{L}\right)L = 2\frac{q}{m}\left(\frac{V_a - V_b}{L}\right)$$

$$K = \frac{1}{2}mv^2 = 2\frac{q}{m}\left(\frac{V_a - V_b}{L}\right)$$

(b) The electric field does the work on the moving charge

$$W = \int_a^b \vec{F}\cdot d\vec{L} = FL = q(EL) = q\left(\frac{V_a - V_b}{L}\right)L$$

$$= q(V_a - V_b).$$

(c) The change in potential is $\Delta V = V_b - V_a$ so that the change in potential energy is

$$\Delta U = q\,\Delta V = q(V_b - V_a).$$

(d) The work done by the field is the negative of the change in potential energy,

$$W = -\Delta U = -q\,\Delta V$$

The work–energy relation is

$$W = \Delta K = -\Delta U$$

or

$$\Delta K + \Delta U = \Delta(K + U) = 0$$

The <u>potential energy plus the kinetic energy is conserved</u>. This implies

$$\Delta E = \frac{1}{2}\,mv^2 + q(V_b - V_a) = 0$$

or

$$\frac{1}{2}\,mv^2 = q(V_a - V_b)$$

as verified above.

Example 5

Find the potential on the x axis at a point P with coordinates (x,0) when two positive point charges of magnitude $Q = 10^{-9}$ C are on the y axis at (0,a) and (0,-a), where a = 1 cm.

     (a) Calculate this by superposition of the potentials due to each charge.

     (b) Calculate this by direct integration of $*\ \vec{E} \cdot d\vec{l}$. Evaluate the results for x = a.

     (c) Find the work done by the electric field of the two charges when a third charge of equal magnitude is moved from the point (2a,0) to (a,0).

Solution:

(a)   Refer to Fig. 23-3:

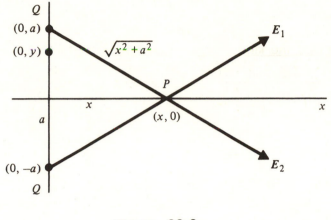

Figure 23-3

The potential at point (x,0) arising from the charges at (0,±a) is the <u>scalar</u> sum of the potentials due to each of the charges:

$$V = \frac{Q}{4\pi\varepsilon_0\sqrt{x^2 + a^2}} + \frac{Q}{4\pi\varepsilon_0\sqrt{x^2 + a^2}}$$

$$V = \frac{2Q}{4\pi\varepsilon_0\sqrt{x^2 + a^2}}$$

(b)   The field **E** on the x-axis at x has only an x component,

$$E_x = E_{1x} + E_{2x} = 2|E_1|\cos\theta$$

$$E_x = 2\ \frac{Q}{4\pi\varepsilon_0(x^2 + a^2)}\ \frac{x}{\sqrt{x^2 + a^2}}$$

$$E_x = \frac{2Qx}{4\pi\varepsilon_0(x^2 + a^2)^{3/2}}$$

43

Thus we have

$$V = \int_x^\infty E_x \cdot dx = \frac{2Q}{4\pi\varepsilon_0} \int_x^\infty \frac{x\, dx}{\left(x^2 + a^2\right)^{3/2}}$$

$$V = \frac{2Q}{4\pi\varepsilon_0} \left[\frac{-1}{\sqrt{x^2 + a^2}} \right]_x^\infty$$

$$V = \frac{2Q}{4\pi\varepsilon_0} \frac{1}{\sqrt{x^2 + a^2}} \qquad \text{as in (a).}$$

At $x = a = 1$ cm $= 10^{-2}$ m,

$$V = \frac{\left(2 \times 10^{-9}\ \text{C}\right)}{(4\pi)\left(8.85 \times 10^{-12}\ \text{C}^2/\text{N}\cdot\text{m}^2\right)\sqrt{2(0.01\ \text{m})^2}}$$

$$= 1270\ \text{V}.$$

(c) $$W = \int_{a'}^b \vec{F} \cdot d\vec{L} = \int_{a'}^b Q\vec{E} \cdot d\vec{L} = Q\left(V_{a'} - V_b\right)$$

Let point a' be at $(0, 2a)$ and point b at $(0, a)$. Then

$$V_a = \frac{2Q}{4\pi\varepsilon_0} \frac{1}{\sqrt{(2a)^2 + a^2}} = \frac{2Q}{4\pi\varepsilon_0 a\sqrt{5}}$$

$$V_b = \frac{2Q}{4\pi\varepsilon_0} \frac{1}{\sqrt{(a)^2 + a^2}} = \frac{2Q}{4\pi\varepsilon_0 a\sqrt{2}}$$

$$W = \frac{2Q^2}{4\pi\varepsilon_0 a} \left[\frac{1}{\sqrt{5}} - \frac{1}{\sqrt{2}} \right]$$

$$W = \frac{2\left(10^{-9}\ \text{C}\right)^2(-\ 0.26)}{(4\pi)\left(8.85 \times 10^{-12}\ \text{C}^2/\text{N}\cdot\text{m}^2\right)(0.01\ \text{m})}$$

$$= -\ 4.68 \times 10^{-7}\ \text{J}.$$

The work done is negative because the displacement is opposite to the direction of the electric force.

Example 6

For the charge distribution of Example 5, Fig. 23-3,
   (a) Find the potential at a point (0,y) on the y-axis between the two charges.
   (b) If an electron is released from rest at the point (0,0) and given a slight nudge, what is its velocity at (0,a/2)?

Solution:

(a) The potential is the scalar sum of the potential arising from each of the charges.

$$V = \frac{Q}{4\pi\varepsilon_0(a - y)} + \frac{Q}{4\pi\varepsilon_0(a + y)}$$

$$V = \frac{2aQ}{4\pi\varepsilon_0\left(a^2 - y^2\right)} .$$

(b) The potential energy of the electron (with charge -e) is

$$U = -eV = \frac{-2aeQ}{4\pi\varepsilon_0\left(a^2 - y^2\right)} .$$

The total energy is conserved,

$$E = K + U$$

$$\Delta E = 0$$

$$\Delta K = \frac{1}{2} mv^2 = -\Delta U = \left[U(0) - U(a/2)\right]$$

Thus the gain in kinetic energy is

$$\Delta K = \frac{-2aeQ}{4\pi\varepsilon_0 a^2} + \frac{2aeQ}{4\pi\varepsilon_0\left(a^2 - (a/2)^2\right)}$$

$$\Delta K = \frac{2aeQ}{4\pi\varepsilon_0 a^2}\left[-1 + \frac{4}{3}\right]$$

$$\Delta K = \frac{eQ}{6\pi\varepsilon_0 a} = \frac{1}{2} mv^2$$

and the velocity is

$$v = \sqrt{\left(\frac{1}{m}\right)\left(\frac{eQ}{3\pi\varepsilon_0 a}\right)}$$

$$v = \sqrt{\frac{(1.6 \times 10^{-19}\,C)(10^{-9}\,C)}{(3\pi)(8.85 \times 10^{-12}\,C^2/\,Nm^2)(9.1 \times 10^{-31}\,kg)(0.01\,m)}}$$

$$= 1.45 \times 10^7\,m/s$$

Example 7

In the previous example, find the field at a point on the y axis and verify that

$$E_y = -\frac{dV}{dy}\,.$$

Solution:

The total field is the sum of the fields arising from each charge,

$$E_y = \frac{-Q}{4\pi\varepsilon_0(a - y)^2} + \frac{Q}{4\pi\varepsilon_0(a + y)^2}$$

This can be verified by calculating the negative y derivative of the potential,

$$-\frac{dV}{dy} = -\frac{d}{dy}\left[\frac{Q}{4\pi\varepsilon_0(a - y)} + \frac{Q}{4\pi\varepsilon_0(a + y)}\right]$$

$$-\frac{dV}{dy} = \frac{-Q}{4\pi\varepsilon_0}\left[\frac{-1}{(a - y)^2}\,(-1) + \frac{-1}{(a + y)^2}\right]$$

$$-\frac{dV}{dy} = \left[\frac{-Q}{4\pi\varepsilon_0(a - y)^2} + \frac{Q}{4\pi\varepsilon_0(a + y)^2}\right] = E_y\,.$$

The result is verified.

Example 8

A coaxial cable consists of an inner solid cylindrical wire conductor of radius $r_1 = 0.2$ mm surrounded by a thin cylindrical coaxial outer sleeve of radius $r_2 = 0.5$ cm. The potential difference between the inner and outer conductors is 220 V. Find the charge per unit area on the inner and outer conductors.

Solution:

The field is radial and depends on the radial coordinate r only. Using Gauss' law on a coaxial cylindrical surface of length L with radius r, $r_1 < r < r_2$,

$$\oint \vec{E} \cdot d\vec{A} = E(2\pi r L) = \frac{Q}{\varepsilon_o} = \frac{\lambda L}{\varepsilon_o}$$

where λ is the charge per unit length on the inner conductor. Thus the field inside the cable is

$$E = \frac{\lambda}{2\pi r \varepsilon_0} = E_r .$$

The potential difference is

$$V_1 - V_2 = \int_1^2 \vec{E} \cdot d\vec{L} = \int_{r_1}^{r_2} E_r \, dr = \int_{r_1}^{r_2} \frac{\lambda}{2\pi \varepsilon_0 r} \, dr$$

$$V_1 - V_2 = \frac{\lambda}{2\pi \varepsilon_0} \ln\left(\frac{r_2}{r_1}\right) = \Delta V .$$

The charge per unit length is

$$\lambda = \frac{2\pi \varepsilon_0 \Delta V}{\ln\left(r_2/r_1\right)} = \frac{2\pi\left(8.85 \times 10^{-12} \ C^2/N\cdot m^2\right)(220 \ V)}{\ln(0.5 \ cm/0.02 \ cm)}$$

$$= 3.80 \times 10^{-9} \ C/m.$$

The charge per unit area is related to the charge per unit length by

$$dq = \sigma \, dA = \lambda \, dl$$

$$\sigma = \lambda \frac{dL}{dA} = \frac{\lambda \, dL}{d(2\pi r L)} = \frac{\lambda}{2\pi r}$$

$$\sigma_1 = \frac{\lambda}{2\pi r_1} = \frac{\left(3.8 \times 10^{-9} \ C/m\right)}{(2\pi)\left(2 \times 10^{-4} \ m\right)}$$

$$\sigma_1 = 3.02 \times 10^{-6} \ C/m^2$$

$$\sigma_2 = \frac{\lambda}{2\pi r_2} = \frac{\left(3.8 \times 10^{-9} \ C/m\right)}{(2\pi)\left(5 \times 10^{-3} \ m\right)}$$

$$= 1.21 \times 10^{-7} \ C/m^2$$

Example 9

As shown in Fig. 23-4, a solid conducting sphere of radius r_a is concentric with a hollow spherical shell of inner radius r_b and outer radius r_c. The potential difference between the spheres is V_{ab}.

    (a) Find the field between the spheres and the charge on the inner sphere.
    (b) Find the charge on the surface at r_b.
    (c) Find the charge on the surface at r_c if $E = 0$ for $r > r_c$.
    (d) Find the field outside *if the outer conductor is neutral, that is has zero net charge.*

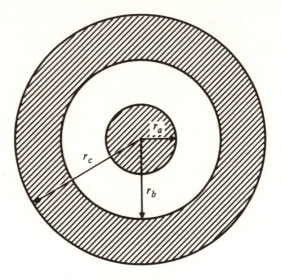

Figure 23-4

Solution:

(a) The field is radial and depends on r only. Let Q be the total charge on the inner sphere, residing on the surface at r_a. We don't know Q, but can find it knowing V_{ab}. By Gauss's law

$$\oint \vec{E} \cdot d\vec{A} = E(4\pi r^2) = \frac{Q}{\varepsilon_o} ; \qquad r_a < r < r_b$$

so that the field in the cavity is

$$E = \frac{Q}{(4\pi r^2)\varepsilon_0} .$$

The potential difference between r_a and r_b is

$$V_{ab} = V_a - V_b = \int_a^b \vec{E} \cdot d\vec{r} = \int_a^b \frac{Q}{4\pi\varepsilon_0} \frac{dr}{r^2} = \frac{Q}{4\pi\varepsilon_0} \left[\frac{-1}{r} \right]_{r_a}^{r_b}$$

$$V_{ab} = \frac{Q}{4\pi\varepsilon_0} \left[\frac{1}{r_a} - \frac{1}{r_b} \right]$$

Thus the charge on the sphere is

$$Q = \frac{4\pi\varepsilon_0 V_{ab}}{\left[\frac{1}{r_a} - \frac{1}{r_b} \right]}$$

and the field is also given by

$$E = \frac{Q}{4\pi\varepsilon_0 r^2} = \frac{V_{ab}}{\left[\frac{1}{r_a} - \frac{1}{r_b} \right] r^2}$$

(b) The field inside the inner sphere and between r_b and r_c is zero because the material is a conductor. By Gauss's law, the net charge enclosed by a sphere of radius r, $r_b < r < r_c$, must be zero. Thus the charge on the surface at r_b is equal and opposite to the charge on the inner sphere,

$$Q = \frac{- 4\pi\varepsilon_0 V_{ab}}{\left[\frac{1}{r_a} - \frac{1}{r_b} \right]}$$

(c) Since the field outside of both conductors is zero, the net charge enclosed in a sphere of radius $r > r_c$ is zero, by Gauss's law. Thus there is no charge on the surface at r_c.

(d) If the outer conductor is neutral, a charge Q must be on the surface at r_c to balance the charge -Q on the surface at r_b. Outside of the system at $r > r_c$ an application of Gauss's law yields

$$\oint \vec{E} \cdot d\vec{A} = \frac{Q - Q + Q}{\varepsilon_0} = \frac{Q}{\varepsilon_0} = E(4\pi r^2)$$

$$E = \frac{Q}{(4\pi r^2)\varepsilon_0} .$$

49

Example 10

Sketch the potential V as a function of r for the system of Example 9 when
(a) the outer conductor has net charge -Q and
(b) the outer conductor is electrically neutral.

Solution:

(a) We choose the reference potential to be $V = 0$ at $r = \infty$. Thus

$$V(r) = \int_r^\infty \vec{E} \cdot d\vec{L}$$

$$V(r) = \int_r^{r_a} E\, dr + \int_{r_a}^{r_b} E\, dr + \int_{r_b}^{r_c} E\, dr + \int_{r_c}^\infty E\, dr \qquad 0 < r < r_a$$

$$V(r) = \int_r^{r_b} E\, dr + \int_{r_b}^{r_c} E\, dr + \int_{r_c}^\infty E\, dr \qquad r_a < r < r_b$$

$$V(r) = \int_r^{r_c} E\, dr + \int_{r_c}^\infty E\, dr \qquad r_b < r < r_c$$

$$V(r) = \int_r^\infty E\, dr \qquad r_c < r$$

The line integral has been split into parts as a convenience because the expressions for the fields may differ in each region. Since $E = 0$ for $r < r_a$ and $r_b < r < r_c$, and $E=0$ for $r > r_c$,

$$V(r) = \int_{r_a}^{r_b} E\, dr \qquad 0 < r < r_a$$

$$V(r) = \int_r^{r_b} E\, dr \qquad r_a < r < r_b$$

$$V(r) = 0, \qquad r_b < r < r_c$$

$$V(r) = 0, \qquad r_c < r$$

Thus

$$V(r) = \int_{r_a}^{r_b} E\,dr = \frac{Q}{4\pi\varepsilon_0}\left[\frac{-1}{r}\right]_{r_a}^{r_b} = \frac{Q}{4\pi\varepsilon_0}\left[\frac{1}{r_a} - \frac{1}{r_b}\right]$$

$$= V_{ab}, \qquad r \leq r_a$$

$$V(r) = \int_{r}^{r_b} E\,dr = \frac{Q}{4\pi\varepsilon_0}\left[\frac{-1}{r}\right]_{r}^{r_b} = \frac{Q}{4\pi\varepsilon_0}\left[\frac{1}{r} - \frac{1}{r_b}\right]$$

$$V(r) = V_{ab}\left[\frac{1}{r_a} - \frac{1}{r_b}\right]^{-1}\left[\frac{1}{r} - \frac{1}{r_b}\right] \qquad r_a < r < r_b$$

$$V = 0 \qquad r > r_b$$

A sketch is given in Fig. 23-5a

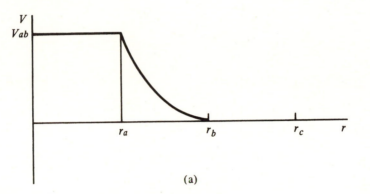

(a)

Figure 23-5a

(b) When the outer conductor is neutral, there is a field outside the system, and

$$V(r) = \int_{r_a}^{r_b} E\,dr + \int_{r_c}^{\infty} E\,dr \qquad 0 < r < r_a$$

$$V(r) = \int_{r}^{r_b} E\,dr + \int_{r_c}^{\infty} E\,dr \qquad r_a < r < r_b$$

$$V(r) = \int_{r_c}^{\infty} E\,dr \qquad r_b < r < r_c$$

$$V(r) = \int_{r}^{\infty} E\,dr \qquad r_c < r$$

Thus

51

$$V(r) = \frac{Q}{4\pi\varepsilon_0}\left[\frac{1}{r_a} - \frac{1}{r_b}\right] + \int_{r_c}^{\infty} \frac{Q}{4\pi\varepsilon_0} \frac{dr}{r^2}$$

$$V(r) = \frac{Q}{4\pi\varepsilon_0}\left[\frac{1}{r_a} - \frac{1}{r_b} + \frac{1}{r_c}\right] \qquad 0 < r < r_a$$

$$V(r) = \frac{Q}{4\pi\varepsilon_0}\left[\frac{1}{r} - \frac{1}{r_b} + \frac{1}{r_c}\right] \qquad r_a < r < r_b$$

$$V(r) = \frac{Q}{4\pi\varepsilon_0}\left[\frac{1}{r_c}\right] \qquad r_b < r < r_c$$

$$V(r) = \frac{Q}{4\pi\varepsilon_0}\left[\frac{1}{r}\right] \qquad r_c < r$$

For a sketch see Fig. 23-5b

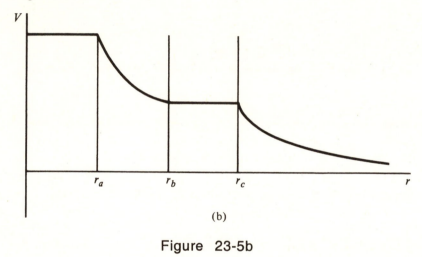

(b)

Figure 23-5b

Example 11

(a) Estimate the energy necessary to separate completely an electron and proton bound by the Coulomb force in a hydrogen atom, normally a distance 10^{-8} cm apart.

(b) Through what voltage must an electron be accelerated to gain this energy?

Solution:

(a) The work done is equal to the increase in potential energy

$$W = U = -eV = -e[V(\infty) - V(r)]$$

52

$$W = -e \left[0 - \frac{e}{4\pi\varepsilon_0 r} \right] = \frac{e^2}{4\pi\varepsilon_0 r}$$

$$W = \frac{\left(1.6 \times 10^{-19} \text{ C}\right)^2}{(4\pi)\left(8.85 \times 10^{-12} \text{ C}^2/\text{N·m}^2\right)\left(10^{-10} \text{ m}\right)}$$

$$W = 2.3 \times 10^{-18} \text{ J}$$

(b) $e\Delta V = 2.3 \times 10^{-18}$ J

$$\Delta V = \frac{2.3 \times 10^{-18} \text{ J}}{1.6 \times 10^{-19} \text{ C}} = 14.4 \text{ V.}$$

QUIZ

1. A positive charge of magnitude 10^{-9} C is released from rest in a constant electric field of magnitude 4×10^4 N/C. What is its kinetic energy when it is 6×10^{-2} m from its initial position?

Answer: 2.4×10^{-6} J

2. Find the potential on the axis of a uniformly charged disk of radius a, a distance x from the disk.

Answer: $V = (\sigma/ 2\varepsilon_0) [(a^2 + x^2)^{1/2} - x]$

3. The conductors of a coaxial conducting cable consist of an inner cylinder of radius one mm and an outer cylinder of radius 5 mm. What is the charge per unit length on the cable when the voltage across the conductors is 8 V?

Answer: 2.8×10^{-10} C/m

4. A pair of charges ± Q are on the x-axis at x = ± s. Find the potential at a point
    (a) on the x-axis
    (b) on the y-axis

Answer: (a) $[Qx/2\pi\varepsilon_0(s^2 - x^2)]$; (b) zero.

5. A spherical conducting ball of radius 10 cm is charged so that it is at a potential of 110 V relative to a distance ground. What is the net charge on the sphere? What potential would be required if the net charge was 1 coulomb?

Answer: 1.22×10^{-9} C; 9×10^{10} V.

24

CAPACITANCE AND DIELECTRICS

OBJECTIVES

In this chapter the idea of *capacitance* between conductors and the effects of *dielectric* materials on electric fields and potentials are developed. Your objectives are to:

Calculate the *capacitance of simple systems* such as parallel plates, concentric spheres and coaxial cylinders.

Calculate the *equivalent capacitance* of networks of capacitors in series and parallel.

Calculate the *energy stored* in a capacitor by finding the work done to charge it.

Describe the effect of *dielectric materials* on capacitance by the *dielectric constant* K.

Interpret this effect as due to *induced charge* Q_i and a *polarization* P.

Define the *displacement* $\vec{D}$ and obtain a Gauss's law for $\vec{D}$.

REVIEW

A capacitor usually consists of two conductors, often with equal and opposite charges, Q, -Q, separated by an insulator. If the potential difference between the conductors is V, the capacitance C is the ratio of charge to potential

$$C = \frac{Q}{V}$$

The unit of capacitance is the (coulomb)/(volt) = *farad* (F).

The capacitance of a parallel plate capacitor is

$$C = \frac{\varepsilon_0 A}{d}$$

where A is the plate area and d the plate separation.

The equivalent capacitances of two capacitors in series or parallel are given in Fig. 24-1.

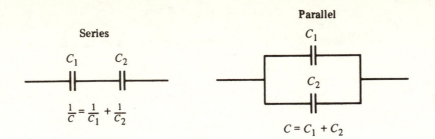

Figure 24-1

The energy needed to place a charge Q on a capacitor is

$$W = \frac{Q^2}{2C} = \frac{1}{2}\left(CV^2\right) = \frac{1}{2}\left(QV\right).$$

This energy can be thought of as stored in the region of space where there is an electric field. The energy per unit volume is

$$u = \frac{1}{2}\left(\varepsilon_0 E^2\right).$$

The capacitance increases when dielectric material is inserted in the region between the plates of a capacitor; that is, for a given potential, the charge on the plates increases when a dielectric is inserted. The dielectric constant of the material may be measured as the ratio of the increased capacitance to the original capacitance:

$$K = \frac{C}{C_0} = \frac{\varepsilon}{\varepsilon_0}$$

where C is the capacitance with dielectric material and C_0 the capacitance in a vacuum. When the charge on the plates is fixed, the effect of the material is to reduce the field (and hence the potential) between the plates for a given charge Q. The capacitance of a parallel plate capacitor with dielectric material between the plates of an area A and separation L is

$$C = KC_0 = K\frac{A\varepsilon_0}{L} = \frac{A\varepsilon}{L}, \qquad \varepsilon = K\varepsilon_0.$$

The electric field is reduced because the molecules of the material line up with the electric field in such a way that part of the field due to the charge Q is canceled by the molecular fields arising from *induced* or bound charge Q_i. This lining up is called polarization, and is measured by the dipole moment per unit volume P. Denoting the integral over a closed surface by: $\oint$; the Gauss's law for P is

$$\oint \vec{P} \cdot d\vec{A} = -Q_i$$

If Q is the charge on conductors ("free charge"), the total charge is $Q + Q_i$ and the Gauss's law for $\mathbf{E}$ is

$$\oint \vec{E} \cdot d\vec{A} = \frac{(Q + Q')}{\varepsilon_0}$$

The vector *electric displacement* $\vec{D}$ is defined as

$$\vec{D} = \varepsilon_0 \vec{E} + \vec{P}$$

and $\vec{D}$ satisfies the Gauss's law

$$\oint \vec{D} \cdot d\vec{A} = Q$$

In dielectrics where P is proportional to E, and has the same direction

$$D = \varepsilon E = K\varepsilon_0 E.$$

HINTS AND PROBLEM-SOLVING STRATEGIES

Please review the two problem solving strategies in the main text, one on equivalent capacitance, and the other on dielectrics. These will be used in the examples below.

QUESTIONS AND ANSWERS

Question. In charging a capacitor, electrons are removed from one plate and transferred to the other. How many electrons must be transferred to create a potential difference of 1 volt across a 1 µF capacitor?

Answer. Since $q = |q_e| = 1.6 \times 10^{-19}$ C, $Nq = C\Delta V = (1\ \mu F)(1\ \text{Volt}) = 10^{-6}$ C. Therefore $N = (10^{-6})/(1.6 \times 10^{-19}) \approx 6 \times 10^{12}$ electrons.

Question. For two parallel plates charged as shown, determine the signs of the particles labeled A, B, and C that are shot between the plates and deflected by the electric field.

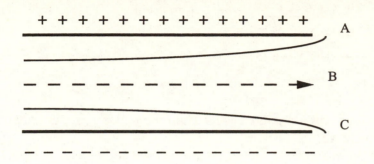

Answer. A is negatively charged, B is neutral or uncharged, and C is positively charged.

EXAMPLES AND SOLUTIONS

Example 1

A parallel plate capacitor has a plate separation of 3 mm and a plate area 100 cm². (Unless otherwise specified, assume that the material between the plates has K=1, $\varepsilon = \varepsilon_0$, true for a vacuum and a good approximation for air.)

      (a) Find its capacitance.
      (b) If the potential across the plates is 110 V, find the charge on each plate.
      (c) Find the field between the plates.
      (d) Find the energy stored in the capacitor.

Solution:

(a) $C = \dfrac{\varepsilon_0 A}{L} = \dfrac{\left(8.85 \times 10^{-12}\ C^2/N \cdot m^2\right)\left(100 \times 10^{-4}\ m^2\right)}{(0.003\ m)}$

       $= 2.95 \times 10^{-11}\ F = 29.5\ pF$

(b) $Q = VC = (110\ V)(2.95 \times 10^{-11}\ F) = 3.25 \times 10^{-9}\ C$

(c) $E = \dfrac{\sigma}{\varepsilon_0} = \dfrac{Q}{A\varepsilon_0} = \dfrac{\left(3.25 \times 10^{-9}\ C\right)}{\left(8.85 \times 10^{-12}\ C^2/N \cdot m^2\right)\left(100 \times 10^{-4}\ m^2\right)}$

       $= 3.67 \times 10^4\ V/m$

Alternatively

$$E = (V/L) = (110 \text{ V}/0.003 \text{ m}) = 3.667 \times 10^4 \text{ V/m}$$

(d) $\quad W = \dfrac{Q^2}{2C} = \dfrac{(3.25 \times 10^{-9} \text{ C})^2}{(2)(2.95 \times 10^{-11} \text{ F})} = 1.79 \times 10^{-7} \text{ J}$

Example 2

A parallel plate capacitor has a charge 10^{-9} C when the plates are 1 cm apart and it is connected to a voltage source of 12 V.

(a) If the plates are pulled to 2 cm separation while keeping the potential constant, what is the new charge?

(b) If the plates are disconnected from the voltage source and pulled to 2 cm separation, while keeping the charge constant, what is the new potential?

Solution:

The key is to recognize the difference between a capacitor connected to a battery (constant voltage) and to a disconnected one (constant charge). The original capacitance is

$$C_1 = \frac{\varepsilon_0 A}{L} = \frac{Q_1}{V_1}$$

The final capacitance is

$$C_2 = \frac{\varepsilon_0 A}{2L} = \frac{Q_2}{V_2} = \frac{C_1}{2}$$

(a) If $V_1 = V_2$ (this is the case if the capacitor is connected to a 12 V battery while the plates are pulled apart),

$$Q_2 = C_2 V_2 = C_2 V_1 = (1/2)(C_1 V_1) = (1/2)(Q_1) = 0.5 \times 10^{-9} \text{ C}.$$

(b) If $Q_1 = Q_2$ (this would be the case if the insulated plates were disconnected from the battery before being pulled apart.)

$$V_2 = (Q_2/C_2) = (2Q_1/C_1) = 2V_1$$

$$= 24 \text{ V}.$$

Example 3

A capacitor with capacitance $C_1 = 30$ μF is charged to a potential of 500 V, disconnected from the source, and then connected in parallel to an uncharged capacitor with capacitance $C_2 = 10$ μF.

(a) Find the charge on the 30 μF capacitor before the connection is made.
(b) Find the final charge on each of the capacitors.
(c) Find the energy lost when the connection is made.
(d) What happens to the energy?

Solution:

(a) Originally we have the charge

$$Q = VC = (500 \text{ V})(30 \times 10^{-6} \text{ F}) = 1.50 \times 10^{-2} \text{ C}$$

(b) When the capacitors are connected in parallel the original charge (left picture) redistributes itself across the two capacitors as shown in Fig. 24-2 (right picture).

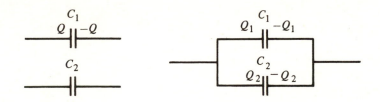

Figure 24-2

$$Q = Q_1 + Q_2$$

In parallel connection, the potential across each capacitor is the same,

$$V_1 = V_2$$

$$\frac{Q_1}{C_1} = \frac{Q_2}{C_2} = \frac{Q - Q_1}{C_2}$$

$$Q_1\left[\frac{1}{C_1} + \frac{1}{C_2}\right] = \frac{Q}{C_2}$$

$$Q_1 = \frac{Q}{C_2} \cdot \left[\frac{1}{C_1} + \frac{1}{C_2}\right]^{-1} = \left(\frac{C_1}{C_1 + C_2}\right)Q$$

Let's approach this problem in another way using the problem-solving strategy of equivalent capacitance.

59

Another way to get this result is to note that the equivalent capacitance of the combination is $C = C_1 + C_2$. The voltage across the system is equal to the voltage across each capacitor,

$$V = \frac{Q}{C} = \frac{Q}{C_1 + C_2} = \frac{Q_1}{C_1}$$

$$Q_1 = \left(\frac{C_1}{C_1 + C_2}\right)Q$$

$$Q_1 = \left(\frac{30 \ \mu F}{30 \ \mu F + 10 \ \mu F}\right)\left(1.5 \times 10^{-2} \ C\right)$$

$$= 1.13 \times 10^{-2} \ C$$

$$Q_2 = Q - Q_1 = 1.50 \times 10^{-2} \ C - 1.13 \times 10^{-2} \ C = 0.37 \times 10^{-2} \ C$$

(c)   The original energy is

$$W = \frac{Q^2}{2C_1} = \frac{\left(1.5 \times 10^{-2} \ C\right)^2}{(2)(30 \times 10^{-6} \ F)} = 3.75 \ J.$$

The final energy is

$$W = \frac{Q_1^2}{2C_1} + \frac{Q_2^2}{2C_2}$$

$$W = \frac{\left(1.13 \times 10^{-2} \ C\right)^2}{(2)(30 \times 10^{-6} \ F)} + \frac{\left(0.37 \times 10^{-2} \ C\right)^2}{(2)(10 \times 10^{-6} \ F)}$$

$$= 2.13 \ J + 0.68 \ J = 2.81 \ J$$

Another way to calculate the energy is to use the equivalent capacitance, $C = C_1 + C_2$, yielding the same result

$$W = \frac{Q^2}{2C} = \frac{\left(1.50 \times 10^{-2} \ C\right)^2}{(2)(40 \times 10^{-6} \ F)} = 2.81 \ J.$$

The energy lost in connecting the capacitors is

$$\Delta W = (3.75 - 2.81)J = 0.94 \ J.$$

(d)   The energy is lost to dissipative processes (like sparks) as the connection is made.

Example 4

Find the capacitance of a capacitor consisting of two thin concentric conducting shells of inner radius $r_a = 1$ cm and outer radius $r_b = 2$ cm.

Solution:

Since capacitance is defined as the ratio of charge (Q) to potential difference (V), we must find the potential difference for a given charge. If Q is the charge on the inner conductor and -Q is the charge on the outer conductor, the field between the shells is

$$E = \frac{Q}{4\pi\varepsilon_0 r^2} .$$

The potential difference between the shells can be calculated from the field,

$$V = V_a - V_b = \int_a^b \vec{E} \cdot d\vec{r} = \int_a^b \frac{Q\, dr}{4\pi\varepsilon_0 r^2} = \frac{Q}{4\pi\varepsilon_0}\left[\frac{1}{r_a} - \frac{1}{r_b}\right]$$

The capacitance is then

$$C = \frac{Q}{V} = (4\pi\varepsilon_0)\cdot\left[\frac{1}{r_a} - \frac{1}{r_b}\right]^{-1}$$

$$C = 4\pi\left(8.85 \times 10^{-12}\ C^2/N\cdot m^2\right)\cdot\left[\frac{1}{0.01\ m} - \frac{1}{0.02\ m}\right]^{-1}$$

$$= 2.22 \times 10^{-12}\ F = 2.22\ pF$$

$$(1\ pF = 10^{-12}\ F\ \text{and}\ 1\ \mu F = 10^{-6}\ F)$$

Example 5

Find the capacitance of two concentric conducting cylinders of inner radius $r_a = 1$ mm and outer radius $r_b = 2$ mm and length $L = 1$ m.

Solution:

The strategy is similar to the last example. The field between the cylinders is

$$E = \frac{\lambda}{2\pi\varepsilon_0 r}$$

where λ is the charge per unit length on the inner conductor. The potential between the conductors is

$$V = \int_a^b \vec{E} \cdot d\vec{r} = \int_{r_a}^{r_b} \frac{\lambda}{2\pi\varepsilon_0} \frac{dr}{r} = \frac{\lambda}{2\pi\varepsilon_0} \ln\left(\frac{r_b}{r_a}\right)$$

The capacitance is

$$C = \frac{Q}{V} = \frac{\lambda L}{(\lambda/2\pi\varepsilon_0)[\ln(r_b/r_a)]} = \frac{2\pi\varepsilon_0 L}{[\ln(r_b/r_a)]} = 8.02 \times 10^{-11} \text{ F} = 80.2 \text{ pF}$$

Example 6

Find the equivalent capacitance between points a and b of the network of Fig. 24-3a.

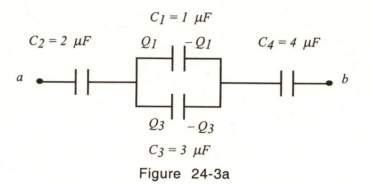

$C_1 = 1 \ \mu F$

$C_2 = 2 \ \mu F$ Q_1 $-Q_1$ $C_4 = 4 \ \mu F$

a b

Q_3 $-Q_3$

$C_3 = 3 \ \mu F$

Figure 24-3a

Solution:

We follow the problem-solving strategy of reducing parts of the network to their replacement equivalents. The 1 μF and 3 μF are in parallel and their equivalent capacitance is 1 μF + 3 μF = 4 μF. This reduces the network to the three series capacitances shown in Fig. 24-3b. The equivalent capacitance of this network is C, with

$$\frac{1}{C} = \frac{1}{C_2} + \frac{1}{C_1 + C_3} + \frac{1}{C_4} = \frac{1}{2 \ \mu F} + \frac{1}{4 \ \mu F} + \frac{1}{4 \ \mu F} = \frac{2 + 1 + 1}{4 \ \mu F}$$

$$= (1/1 \ \mu F) \quad C = 1 \ \mu F$$

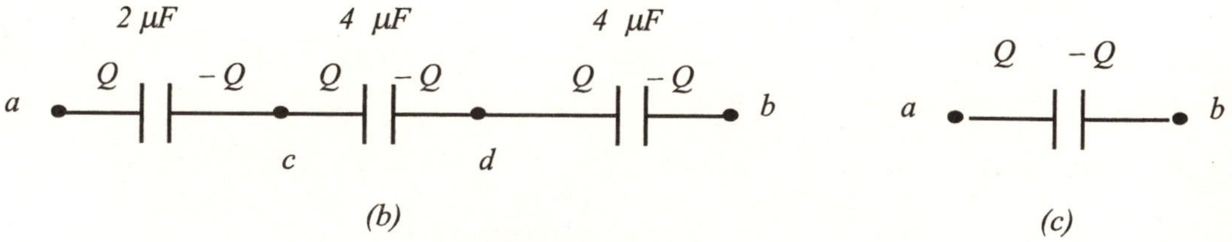

$2 \ \mu F$ $4 \ \mu F$ $4 \ \mu F$ Q $-Q$

a Q $-Q$ Q $-Q$ Q $-Q$ b a b

     c d

(b)   (c)

Figure 24-3 b and c

Example 7

In the network of Fig. 24-3, $V_{ab} = 1000$ V. Find the charges on each capacitor and the potential drops V_{ac}, V_{cd}, and V_{db}.

Solution:

The charge on the equivalent capacitor (see Fig. 24-3c) is

$$Q = VC = (1000 \text{ V})(1 \text{ } \mu\text{F}) = 10^{-3} \text{ C.}$$

The charge on each of the series reduced capacitors is the same, as shown in Fig. 24-3b. The potential drops are thus

$$V_{ac} = (Q/C_{ac}) = (10^{-3} \text{ C}/2 \text{ } \mu\text{F}) = 500 \text{ V}$$

$$V_{cd} = (Q/C_{cd}) = (10^{-3} \text{ C}/4 \text{ } \mu\text{F}) = 250 \text{ V}$$

$$V_{db} = (Q/{}_iC_{db}) = (10^{-3} \text{ C}/4 \text{ } \mu\text{F}) = 250 \text{ V}$$

Note $V_{ac} + V_{cd} + V_{db} = V_{ab}$.

Now we can find the charge on the 1 μF and 3 μF capacitors. Since they are in parallel they have a common potential drop V_{cd}. Thus

$$Q_1 = C_1 V_{cd} = (1 \text{ } \mu\text{F})(250 \text{ V}) = 0.25 \times 10^{-3} \text{ C}$$

$$Q_3 = C_3 V_{cd} = (3 \text{ } \mu\text{F})(250 \text{ V}) = 0.75 \times 10^{-3} \text{ C}$$

Note

$$Q_1 + Q_3 = 10^{-3} \text{ C} = Q.$$

Example 8

Suppose now that the charged capacitors C_1 and C_3 of Fig. 24-3a, Examples 6 and 7, are disconnected and reconnected with terminals of unlike sign together. Find the new charge on each capacitor and the potential across each.

Solution:

This problem sounds complicated but simplifies when we break it down into small steps. The key is to recognize that after the re-connection, the charges must re-distribute to make sure the potential difference across each is the same, because they are in parallel. We originally have the configuration of Fig. 24-4a and then for a moment the configuration of Fig. 24-4b after the crossed re-connection; the charges then redistribute so that the potential across each capacitor is equal, as shown in Fig. 24-4c. Since total charge is conserved,

$$Q_1' + Q_3' = Q_1 - Q_3 = (0.25 - 0.75)10^{-3} \text{ C}$$

$$= -0.50 \times 10^{-3} \text{ C}$$

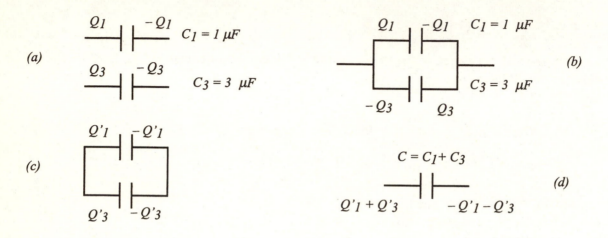

Figure 24-4

The potential across the combination may be calculated by noting that the equivalent capacitor shown in Fig. 24-4d has capacitance,

$$C = C_1 + C_3 = 4 \text{ }\mu\text{F}$$

so that the potential across it is

$$V = (Q/C) = (0.50 \times 10^{-3} \text{ C}/4 \text{ }\mu\text{F}) = 125 \text{ V}.$$

Since $Q_1' + Q_3'$ is negative the potential across the combination is $V' = -125$ V. The charge on each capacitor is

$$Q_1' = V'C_1 = -125 \text{ V } (1 \text{ }\mu\text{F}) = -125 \times 10^{-6} \text{ C}$$

$$Q_3' = V'C_3 = -125 \text{ V } (3 \text{ }\mu\text{F}) = -375 \times 10^{-6} \text{ C}$$

Note

$$Q_1' + Q_3' = -500 \times 10^{-6} \text{ C as calculated above.}$$

64

Example 9

A parallel plate capacitor of area A has charge Q and separation x.

(a) Find the potential energy stored in the field in terms of Q, A and x.

(b) Find the force of attraction between the plates by calculating the field at one plate arising from the charge on the other plate.

(c) Find the work necessary to separate the plates an additional distance Δx in terms of the force F and Δx. By equating this to the increase in potential energy, find F, checking the results of (b).

Solution:

(a) The potential energy is

$$W = \frac{1}{2} QV = \frac{1}{2} \frac{Q^2}{C} ; \quad C = \frac{A\varepsilon_0}{x}$$

$$W = \frac{1}{2} \frac{Q^2 x}{A\varepsilon_0} .$$

(b) The field *due only to one plate* is half the total field,

$$E' = \frac{\sigma}{2\varepsilon_0} = \frac{Q}{2A\varepsilon_0}$$

The force *of one plate* on the *other* is the product of the field due to one plate and the charge on the other plate,

$$F = QE' = \frac{Q^2}{2A\varepsilon_0}$$

(c) The work done by an external force when the plates are separated a distance Δx is

$$F\Delta x = \Delta W = \frac{1}{2} \frac{Q^2 \Delta x}{A\varepsilon_0}$$

$$F = \frac{Q^2}{2A\varepsilon_0}$$

Example 10

A parallel plate capacitor has a capacitance of 1 pF = 10^{-12} F and is filled with dielectric material with a dielectric coefficient K = 3. The dielectric breaks down and becomes conducting at an electric field of 2 x 10^5 V/cm.

(a) If the plate separation is 1 mm, what is the area of the capacitor?

(b) What is the maximum potential that may be put across the capacitor before it breaks down?

Solution:

Now is a good time to review the dielectrics problem-solving strategy of the main text.

(a) The capacitance of the parallel plate capacitor when filled with dielectric material is

$$C = KC_0 = K\frac{A\varepsilon_0}{L}$$

Thus

$$A = \frac{LC}{K\varepsilon_0} = \frac{\left(10^{-3}\ m\right)\left(10^{-12}\ F\right)}{(3)\left(8.85 \times 10^{-12}\ C^2/N{\cdot}m^2\right)}$$

$$= 3.77 \times 10^{-5}\ m^2 = 0.37\ cm^2$$

(b) The field between the plates is

$$E = \frac{V}{L} \quad \text{so that} \quad E_{max} = \frac{V_{max}}{L}$$

$$V_{max} = LE_{max} = (10^{-3}\ m)(2 \times 10^5\ V/cm)$$

$$= 2 \times 10^4\ V$$

Example 11

A parallel plate capacitor of area A and plate separation L is filled with a removable dielectric slab of dielectric constant K. The capacitor is given a charge Q with the slab removed, disconnected from the battery, and then the slab is inserted.

(a) Find the potential difference without the slab.

(b) Find the potential difference with the slab.

(c) Find the field between the plates without the slab.

(d) Find the field with the slab.

(e) Find the induced or bound charge Q_i on the dielectric.

(f) Find the displacement D with and without the slab.

(g) Find the polarization P of the dielectric slab.

66

Solution:

There are a lot of questions here--but we approach this step-by-step.

(a) $Q = VC$ $\qquad\qquad\qquad$ $V = (Q/C) = (QL/A\varepsilon_0)$ $\quad$ (no slab)

(b) $V' = (Q/C') = (QL/A\varepsilon)$ $\qquad$ (slab in, capacitance changes)

$$V' = \frac{QL}{KA\varepsilon_0} = \frac{V}{K} < V \qquad\qquad \text{(voltage changes, charge the same)}$$

(c) $E = (V/L) = (Q/A\varepsilon_0)$

(d) $E' = \dfrac{V'}{L} = \dfrac{Q}{A\varepsilon} = \dfrac{Q}{KA\varepsilon_0} = \dfrac{E}{K} < E$ $\quad$ (new field is reduced by induced charge Q_i)

(e) $E' = \dfrac{Q + Q_i}{A\varepsilon_0} = \dfrac{Q}{A\varepsilon}$

$$Q + Q_i = \frac{\varepsilon_0}{\varepsilon} Q = \frac{Q}{K}$$

$$Q_i = Q\left[\frac{1}{K} - 1\right]$$

Note the bound charge on the slab surface is opposite in sign to the contiguous free charge on the plate, because $K > 1$.

(f) $D = \varepsilon E = \varepsilon_0 E = (Q/A)$ $\qquad$ (no slab)

$$D' = \varepsilon E = \frac{\varepsilon Q}{A\varepsilon_0} = \frac{KQ}{A} \qquad \text{(slab)}$$

(g) $P = D - \varepsilon_0 E = 0$ $\qquad$ (no slab)

$$P = \varepsilon E - \varepsilon_0 E = (\varepsilon - \varepsilon_0)(Q/A\,\varepsilon_0) \qquad \text{(slab)}$$
$$= (K - 1)(Q/A)$$

Example 12

In Example11, assume the capacitor is always connected to a battery keeping it at a voltage V.

  (a) Find the charge on the plates *without* the slab.
  (b) Find the charge on the plates *with* the slab inserted.
  (c) Find the field without the slab.
  (d) Find the field with the slab.
  (e) Find the bound or induced charge Q_i when the slab is inserted.
  (f) Find the displacement vector with and without the slab.
  (g) Find the polarization of the dielectric slab.

Solution:

Again, we approach in small steps.

(a) $Q = VC = (VA\varepsilon_0/L)$

(b) $Q' = VC' = (VA\varepsilon/L) > Q$ (increases)

(c) $E = (V/L)$

(d) $E' = (V/L)$ (same)

(e) $E = \dfrac{Q_{total}}{A\varepsilon_0} = \dfrac{Q}{A\varepsilon_0} = \dfrac{V}{L}$ (without slab)

   $E' = \dfrac{Q_{total}}{A\varepsilon_0} = \dfrac{Q' + Q_i'}{A\varepsilon_0}$ (with slab)

Equating the two fields by (c) and (d), and using the results of (a) and (b), we have

$$\frac{Q + Q'_i}{\varepsilon_0} = \frac{VA}{L} = \frac{VA\varepsilon/L + Q'_i}{\varepsilon_0}$$

Solving for the induced charge, Q'_i,

$$Q_i = \left(\varepsilon_0 - \varepsilon\right)\left(\frac{VA}{L}\right)$$

(f) $D = \varepsilon_0 E = \dfrac{\varepsilon_0 V}{L} = \dfrac{Q}{A}$ (without slab)

   $D = \varepsilon E = \dfrac{\varepsilon V}{L} = \dfrac{Q'}{A}$ (slab)

(g) $P = (\varepsilon - \varepsilon_0) E = (\varepsilon - \varepsilon_0)\dfrac{V}{L}$

QUIZ

1. A parallel plate capacitor has a plate area of 4×10^{-2} m^2 and a plate separation of 10^{-2} m. The potential difference across the plates is 25 V.

(a) Find the capacitance of the capacitor.
(b) Find the charge on the plates.
(c) Find the electric field between the plates.
(d) Find the energy stored in the capacitor.

Answer: 35.4×10^{-12} F, 0.88×10^{-9} C, 2500 V/m, 1.1×10^{-8} J

2. A 1 μF and a 2 μF capacitor are connected in parallel across a 600 V line.

(a) Find the charge on each capacitor and the voltage across each.
(b) The charged capacitors are then disconnected from the line and each other, and reconnected with the terminals of unlike sign together. Find the final charge on each and the voltage across each.

Answer: (a) 1.2 mC, 0.6 mC, V = 600 V.   (b) 200 V, 0.2 mC, 0.4 mC

3. An air-gap capacitor remains connected to a battery. How do the following change when a dielectric slab of constant K = 2 is inserted between the plates:

(a) electric field.
(b) charge on the plates.
(c) potential across the plates.

Answer: (a) remains the same   (b) doubles   (c) remains the same

4. An air-gap capacitor is given a charge Q and then disconnected from the battery. How do the following change when a dielectric slab of constant K = 2 is inserted between the plates:

(a) electric field.
(b) charge on the plates.
(c) potential across the plates.

Answer: (a) halves   (b) remains the same   (c) halves

5. What is the capacitance per unit length of a coaxial cable consisting of an inner conducting sleeve of radius 2 mm and an outer conducting sheath of radius 6 mm?

Answer: C/L = 5.1×10^{-11} F/m.

25
CURRENT, RESISTANCE AND ELECTROMOTIVE FORCE

OBJECTIVES

In this chapter you will be introduced to the basic elements of circuits: current, resistance, potential differences, emf's and power dissipated. Your objectives are to:

Define *electric current* I and electric *current density* J. Relate them to the density and velocities of moving charge carriers.

Apply *Ohm's law* for I and J, in terms of *resistivity* ρ and *resistance* R. Note change of resistance with temperature.

Define an *electromotive force* (emf). Find the open and closed circuit potential difference across a battery of given emf and internal resistance.

Find potential drops around a simple circuit with both emf's and resistances.

Develop and apply the idea of *work and power* in elements of a simple circuit.

Gain a picture of how and *why metals conduct* electric current.

REVIEW

When a charge flows through an area there is a *current*

$$I = \frac{dQ}{dt} = \sum_i n_i q_i v_i A$$

where

n_i = particles of type i per unit volume

q_i = charge on particles of type i

v_i = velocity of particles of type i

A = cross section area through which charge flows.

The unit of current I is coulomb per second = ampere (A).

The current per unit area is the *current density*.

$$J = \frac{I}{A} = \frac{1}{A}\frac{dQ}{dt} = \sum_i n_i q_i v_i$$

The unit of current density is ampere per square meter = A/m^2.

In many conductors the current density J is proportional to the electric field E. The constant of proportionality is the *resistivity* ρ,

$$\rho = \frac{E}{J} = \frac{EA}{I} = \frac{VA}{LI}$$

where V is the potential drop along a field line of length L. The units of ρ are (volt·meter)/(ampere) = ohm·meter (Ω·m). The potential drop is

$$V = I\left(\frac{L\rho}{A}\right) = IR$$

where the resistance R of a wire of length L and cross section A is

$$R = \left(\frac{L\rho}{A}\right).$$

$E = \sigma J$ and $V = IR$ are two forms of *Ohm's* law. The unit of resistance is volt/(ampere) = ohm (Ω).

The resistivity and resistance of metals increase with temperature; for a temperature around some reference temperature T_0, the change is described by

$$\rho_T = \rho_0[1 + \alpha(T - T_0)]$$
$$R_T = R_0[1 + \alpha(T - T_0)]$$

where α is the temperature coefficient of resistivity.

To maintain a steady current in a conductor with resistance, one requires a closed circuit containing an emf (electromotive force) which maintains the potential difference across the resistance. A battery may produce the emf. (See Fig. 25-1).

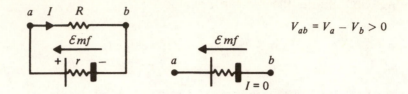

$$V_{ab} = V_a - V_b > 0$$

Figure 25-1

The battery has a positive and negative terminal, and may have an internal resistance r. The internal and external resistances are indicated by the sawtooth symbol in the diagram in Fig. 25-1a. The electromotive force is in the direction indicated, raising the potential from V_b to V_a when crossing from the negative to the positive terminal. The potential then drops as the current flows along the resistor, in the direction of the electric field. The potential rise across the emf is equal to the potential drop across the resistor R. The emf of such a battery is the "open circuit" potential difference

$$V_{ab} = emf(\mathcal{E}).$$

If the emf has no internal resistance then, when connected as in Fig. 25-1a,

$$\mathcal{E} = I\,r \qquad\qquad I = (\mathcal{E}/R)$$

Otherwise, the internal resistance also contributes to the potential drop:

$$\mathcal{E} = IR + Ir$$

$$I = (\mathcal{E}/R + r)$$

$$V_{ab} = \mathcal{E} - Ir = IR.$$

If the battery is "short circuited" R = 0 and

$$I = (\mathcal{E}/r)$$

$$V_{ab} = \mathcal{E} - Ir = 0$$

All these relations are summarized in the Kirchhoff's loop rule,

$$\Sigma\,\mathcal{E} - \Sigma\,IR = 0 \qquad \text{(closed circuit)}$$

which states that the sum of all the emf's about a closed circuit minus the sum of all the potential drops, IR, is zero.

72

The power P necessary to maintain a current I across a potential difference V_{ab} is

$$P = V_{ab} I = I^2R = (V^2/R)$$

If the potential difference is produced by a battery, the power delivered to the external resistance R is

$$P = (\mathcal{E} - Ir)I = \mathcal{E}I - I^2r.$$

Since

$$\mathcal{E}I = I^2R + I^2r$$

the power developed by a battery of emf ($\mathcal{E}$) goes partly into the external circuit (I^2R) and is partly lost to the internal resistance (I^2r).

HINTS and PROBLEM-SOLVING STRATEGIES

Review the problem-solving strategy for power and energy in circuits in the main text. When finding voltage drops around a simple circuit:

(1) Decide on a positive direction of the circuit. (Clockwise or counterclockwise).

(2) Note $\mathcal{E}$ is positive if its open circuit voltage increases along this direction.

(3) Note I is positive if I points in this direction.

QUESTIONS AND ANSWERS

Question. The resistance of a wire changes with temperature. Is this effect due to thermal expansion of the wire?

Answer. No. The material dependent property, the resistivity, which has no factors of length, area, etc. Changes with temperature. The temperature coefficient of resistivity can be one hundred times larger than the thermal expansion coefficient.

Question. Consider two light bulbs, one rated at 50 W, 120 V and the other rated at 60 W, 120 V. What is the ratio of their resistances?

Answer. Since power $P = V^2/R$, we can write: R(50 W)/R(60 W) = 6/5 = 1.2 since the voltage drops are the same.

EXAMPLES AND SOLUTIONS

Example 1

Find the resistance of a 10 cm long aluminum bar of cross section area 1 cm². (See table 26-1 in the text.)

Solution:

$$R = \left(\frac{\rho L}{A}\right) = \frac{(2.63 \times 10^{-8} \ \Omega \cdot m)(0.1 \ m)}{(10^{-2} \ m^2)}$$

$$= 2.63 \times 10^{-5} \ \Omega$$

Example 2

An antique vacuum tube carries an electron current of 10 mA across a flow cross section of area 1 cm². Find the density of electrons at a point in the gap where their velocity is 10^8 m/s.

Solution:

$$J = \frac{I}{A} = nve$$

$$n = \frac{I}{Ave} = \frac{(10 \times 10^{-3} \ A)}{(10^{-2} \ m^2)(10^8 \ m/s)(1.6 \times 10^{-19} \ C)}$$

$$= 6.25 \times 10^{12} \ m^{-3}$$

Example 3

A copper wire of cross section area 5 (mm)² carries a current of 5 A. The density of free electrons is 10^{29} m⁻³.

    (a) How many electrons pass through a cross section of the wire per unit time?
    (b) What is the current density in the wire?
    (c) What is the drift velocity of the electrons?
    (d) If the wire is 1 m long, what is its resistance?
    (e) What is the electric field in the wire?
    (f) What is the potential drop along the wire. Verify Ohm's law.

Solution:

(a) If the number of conducting electrons is N, the charge is Q = Ne and the current is

$$I = \frac{dQ}{dt} = e \frac{dN}{dt}$$

The rate at which electrons pass through the wire is

$$\frac{dN}{dt} = \frac{I}{e} = \frac{(5 \text{ A})}{(1.6 \times 10^{-19} \text{ C})}$$

$$= 3.13 \times 10^{19} \text{ electrons per second.}$$

(b) $J = (I/A) = [(5 \text{ A})/(5 \times 10^{-6} \text{ m}^2)] = 10^6 \text{ A·m}^{-2}$

(c) $J = nve$

$$V = \frac{J}{ne} = \frac{(10^6 \text{ A/m}^2)}{(10^{29} \text{ m}^{-3})(1.6 \times 10^{-19} \text{ C})}$$

$$= 6.25 \times 10^{-5} \text{ m/s.}$$

(d) The resistivity of copper (see table 26-1 in the text) is

$$\rho = 1.72 \times 10^{-8} \text{ }\Omega\text{·m} = 1.72 \times 10^{-8} \text{ V·m/A}$$

$$R = \frac{\rho L}{A} = \frac{(1.72 \times 10^{-8} \text{ }\Omega\text{·m})(1 \text{ m})}{5(10^{-3})^2}$$

$$= 3.44 \times 10^{-3} \text{ }\Omega.$$

(e) $E = \sigma J = (1.72 \times 10^{-8} \text{ V·m/A})(10^6 \text{ A/m}^2)$

$$= 1.72 \times 10^{-2} \text{ V/m.}$$

(f) $V = EL = (1.72 \times 10^{-2} \text{ V/m})(1 \text{ m}) = 1.72 \times 10^{-2} \text{ V}$

$IR = (5 \text{ A})(3.44 \times 10^{-3} \text{ }\Omega) = 1.72 \times 10^{-2} \text{ V}$

$V = IR$ verified

Example 4

The current in a wire varies with time according to

$$I = I_0 e^{-(t/\tau)} \quad \text{where } \tau = 10^{-6} \text{ s and } I_0 = 2 \text{ A}$$

Find the total charge which passes through the wire
  (a) between $t = 0$ and $t = \tau$
  (b) between $t = 0$ and $t \gg \tau$.

Solution:

(a) $I = I_0 e^{-t/\tau} = (dQ/dt)$

$$Q = \int_0^Q dQ = \int_0^t I_0 e^{-(t/\tau)} dt$$

$$Q = I_0 \left[\frac{e^{-(t/\tau)}}{-\tau^{-1}} \right]_0^t = I_0 \tau \left(1 - e^{-1}\right)$$

$$= (2 \text{ A})(10^{-6} \text{ s})(0.63) \quad = 1.26 \times 10^{-6} \text{ C}$$

(b)

$$Q = I_0 \left[\frac{e^{-(t/\tau)}}{-\tau^{-1}} \right]_0^\infty = I_0 \tau = 2 \times 10^{-6} \text{ C}$$

Example 5

A copper wire has a resistance of 10^{-2} Ω at 20° C. What is its resistance at 100° C?

Solution:

The resistivity varies with temperature T according to

$$\rho = \rho_0[1 + \alpha(T - T_0)]$$

where

ρ_0 = resistivity at 20° C and $T_0 = 20°$ C

$\alpha = 0.00393 \,^{\circ}C^{-1}$ (Table 25-2, in the text)

The resistance of the wire is

$$R = \frac{\rho L}{A} = \frac{\rho_0 L}{A}\left[1 + \alpha(T - T_0)\right]$$

$$= R_0[1 + \alpha(T - T_0)]$$

where R_0 is the resistance at $T_0 = 20^{\circ}$ C. Thus at $T = 100^{\circ}$ C the resistance is

$$R = 10^{-2}\,\Omega[1 + 0.00393^{\circ}\ C^{-1}(100^{\circ}\ C - 20^{\circ}\ C)] = 1.31 \times 10^{-2}\,\Omega$$

Example 6

A parallel plate capacitor whose capacitance in air is C_0 is filled with a dielectric of constant K and resistivity ρ. Find the leakage current in terms of the charge Q on the capacitor, ρ, ε_0, and K.

Solution:

The field between the plates is

$$E = \frac{V}{L} = \rho J = \frac{\rho I}{A}$$

where L is the plate separation and A the plate area. The current I is thus

$$I = \frac{AV}{\rho L} = \frac{KA\varepsilon_0}{L}\frac{V}{\rho \varepsilon_0 K} = \frac{KC_0 V}{\rho \varepsilon_0 K} = \frac{C_0 V}{\rho \varepsilon_0 K}$$

$$I = \frac{Q}{\rho \varepsilon_0 K}$$

Example 7

A battery has an open circuit potential difference of 3.0 V and a short-circuit current of 10 A. Find its internal resistance.

Solution:

The open circuit potential difference is the emf $\mathcal{E} = 3.0$ V. In the short circuit condition the only resistance is the internal resistance:

$$\Sigma \mathcal{E} = \Sigma IR$$

$$\mathcal{E} = Ir$$

$$r = (\mathcal{E}/I) = (3.0 \text{ V}/10 \text{ A}) = 0.3 \; \Omega$$

Example 8

If the battery of Example 7 is connected to an external resistance R and the current is 1 A, what is the external resistance?

Solution:

$$\Sigma \mathcal{E} = \Sigma IR$$

$$\mathcal{E} = Ir + IR$$

$$\frac{\mathcal{E} - Ir}{I} = R = \frac{(3.0 \text{ V}) - (1 \text{ A})(0.3 \; \Omega)}{(1 \text{ A})}$$

$$R = 2.7 \; \Omega$$

Example 9

(a) Find the current in the circuit of Fig. 25-2.

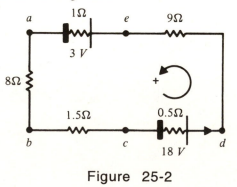

Figure 25-2

(b) Find the potential differences V_{ab}, V_{bc}, V_{cd}, V_{de}, and V_{ea}. Verify that they sum to zero around the closed loop.

Solution:

(a) $\Sigma \mathcal{E} = \Sigma$ IR is the statement that the algebraic (signs important) voltage increases due to emf's is balanced by the voltage drops across the resistances. To use this expression a positive *circuit direction* must first be chosen. This has been taken to be counterclockwise in Fig. 25-2. Then a direction of the current must be assumed. This also has been taken counterclockwise in Fig. 25-2. Emf's are taken positive in the loop rule if they point in the positive circuit direction, negative otherwise. The potential drop IR is positive if I points in the positive direction, negative otherwise. In Fig. 25-2, the 3 V emf is a negative one and the 18 V emf is a positive one. All IR drops are positive.

$$\Sigma \mathcal{E} = 18 \text{ V} - 3 \text{ V} = 15 \text{ V}$$

$$\Sigma \text{ IR} = I(1 \text{ } \Omega + 9 \text{ } \Omega + 0.5 \text{ } \Omega + 1.5 \text{ } \Omega + 8 \text{ } \Omega) = (I)(20 \text{ } \Omega)$$

Thus

$$15 \text{ V} = (I)(20 \text{ } \Omega)$$

$$I = (15 \text{ V}/20 \text{ } \Omega) = 0.75 \text{ A}$$

(b) $\quad V_{ab} = (0.75 \text{ A})(8 \text{ } \Omega) = 6 \text{ V}$

$\quad V_{bc} = (0.75 \text{ A})(1.5 \text{ } \Omega) \text{ } 1.13 \text{ V}$

$\quad V_{cd} = Ir - \mathcal{E} = (0.5 \text{ } \Omega)(0.75 \text{ A}) - 18 \text{ V} = -17.63 \text{ V}$

(The battery tends to make the potential higher at d than c; hence its emf contributes negatively to $V_{cd} = V_c - V_d$)

$\quad V_{de} = (0.75 \text{ A})(9 \text{ } \Omega) = 6.75 \text{ V}$

$\quad V_{ea} = Ir + \mathcal{E}$

$$= (0.75 \text{ A})(1 \text{ } \Omega) + 3 \text{ V} = 3.75 \text{ V}$$

(The 3 V battery is said to be charging because the current is passing through it opposite in direction to the emf.) It may be readily verified that

$$V_{ab} + V_{bc} + V_{cd} + V_{de} + V_{ea} = 0 = V_{aa}.$$

The sum of the potential differences around any closed loop is zero.

Example 10

A 6 V battery has an internal resistance of 0.5 Ω and is connected to a 5.5 Ω resistance.
    (a) What is the current in the circuit when the circuit is closed ?
    (b) What is the potential difference across the battery terminals when the circuit is closed ?
    (c) What is the potential difference across the battery when the circuit is open ?

Solution:

The "internal resistance" is like any other resistance in the circuit.

(a) $\Sigma \mathcal{E} = \Sigma IR$

        $= Ir + IR = I(6\ \Omega)$

      $I = (6\ V/6\ \Omega) = 1\ A$

(b) $V = \mathcal{E} - Ir = 6\ V - (1\ A)(0.5\ \Omega) = 5.5\ V$

(c) $V = \mathcal{E} = 6\ V$

Example 11

    (a) What is the resistance of a 60 W light bulb designed for use on 120 V lines?
    (b) What power would it draw if it were operated on a 240 V line?

Solution:

(a) $P = IV = (V^2/R) = [(120\ V)^2/R] = 60\ W$

      $R = [(120\ V)^2/60\ W] = 240\ \Omega$

(b) $P = (V^2/R) = [(240\ V)^2/240\ \Omega] = 240\ W$

(The bulb would quickly overheat and burn out.)

Example 12

A 12 V battery with internal resistance 0.3 Ω is discharging through a 11.7 Ω resistor? What power is dissipated in the 11.7 Ω resistor?

Solution:

$$\mathcal{E} = I(r + R)$$

$$I = \frac{\mathcal{E}}{r + R} = \frac{(12\ V)}{(0.3\ \Omega + 11.7\ \Omega)} = \frac{(12\ V)}{(12\ \Omega)} = 1\ A$$

$$P = VI = I^2R = (1\ A)^2(11.7\ \Omega) = 11.7\ W$$

Example 13

If the battery in Example 12 is charged by connecting it to a 24 V source, what power does the source develop?

Solution:

Referring to Fig. 25-3,

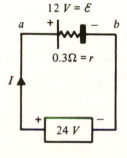

Figure 25-3

we see that the potential difference between a and b is:

$$V_{ab} = Ir + \mathcal{E}$$

$$I = \frac{V_{ab} - \mathcal{E}}{r} = \frac{(24\ V) - (12\ V)}{(0.3\ \Omega)} = 40\ A$$

$$P = V_{ab}I = (24\ V)(40\ A) = 960\ W$$

(In the discharging configuration, the current reverses and $V_{ab} = \mathcal{E} - Ir$.)

QUIZ

1. An aluminum wire is one quarter as long as a copper wire of diameter 2 mm. If the aluminum wire has the same resistance as the copper wire, what is its diameter?

Answer: 1.24 mm

2. A battery has an emf of 12 V and an internal resistance of 2 Ω.
    (a) What is the open circuit potential difference across the terminals of the battery
    (b) What is the current in the battery when it is short circuited?
    (c) What is the current when the battery is connected to an external resistance of 4 Ω?
    (d) What is the potential difference across the terminals of the battery under the conditions of (c)?

Answer: 12 V, 6 A, 2 A, 8 V

3. A flashlight consists of two 1.5 V batteries connected in series to a bulb with resistance 15 Ω.
    (a) What is the internal resistance of *each* battery when the power delivered to the bulb is 0.6 W?
    (b) What is the internal resistance of *each* battery when the power delivered to the bulb is 0.3 W?

Answer: (a) zero   (b) 3.1 Ω

4. 25% of the power developed by a battery in a circuit is lost to internal resistance heating. What is the ratio of the internal to the external resistance?

Answer: 1:3

5. A 9 V battery with internal resistance 0.2 Ω is charged by connecting it to an 18 V source. What power does the source develop?

Answer: 810 W.

26
DIRECT-CURRENT CIRCUITS

OBJECTIVES

The circuit rules of the previous chapter are generalized to include networks with branches and more than one current loop. Your objectives are to:

Find the *equivalent resistance* of resistors in series and parallel.

Apply *Kirchhoff's rules* to circuits with more than one loop, including electrical instruments for measuring potential, current, and resistance.

Design an *ammeter* and *voltmeter* for a given full scale deflection, given a galvanometer and its full scale deflection current.

Find how charges, currents and potentials change with time in an *R-C circuit*.

Gain an understanding of grounded parallel wiring for household and automobile *electric power distribution*.

REVIEW

If resistances R_1 and R_2 are connected in series, their equivalent resistance R is given by

$$R = R_1 + R_2. \qquad \text{(series)}$$

If the same resistances are connected in parallel, their equivalent resistance is given by

$$\frac{1}{R} = \frac{1}{R_1} + \frac{1}{R_2} \qquad \text{(parallel)}$$

Kirchhoff's rules are:

(1) The algebraic sum of the currents *toward* any branch point is zero.

$$\Sigma I = 0.$$

A current is counted positive if it flows into the branch point and negative if it flows out of the branch point. For examples, see Fig. 26-1.

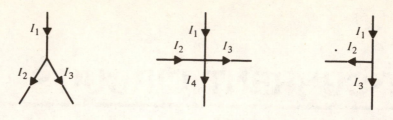

Figure 26-1

$$I_1 - I_2 - I_3 = 0 \quad I_1 + I_2 - I_3 - I_4 = 0 \quad I_1 - I_2 - I_3 = 0$$

$$I_1 = I_2 + I_3 \qquad I_1 + I_2 = I_3 + I_4 \qquad I_1 = I_2 + I_3$$

(2) The algebraic sum of all the potential differences in any loop, including those associated with emf's and resistances, is zero. When IR is a potential drop, it counts as a negative potential difference.

$$\Sigma \mathcal{E} - \Sigma IR = 0$$

$$\Sigma \mathcal{E} = \Sigma IR$$

As in the last chapter, an emf is counted positive if it points in the loop direction and negative otherwise. I is taken positive in the rule if it points in the circuit direction, negative otherwise. For examples see Fig. 26-2, where the arbitrarily chosen positive direction is indicated in each loop by the circulating arrow.

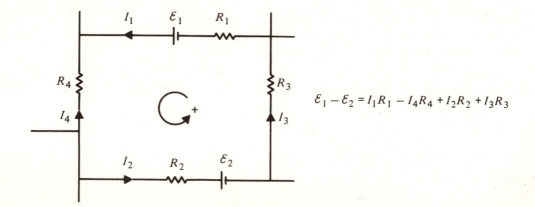

$$\mathcal{E}_1 - \mathcal{E}_2 = I_1 R_1 - I_4 R_4 + I_2 R_2 + I_3 R_3$$

Figure 26-2a

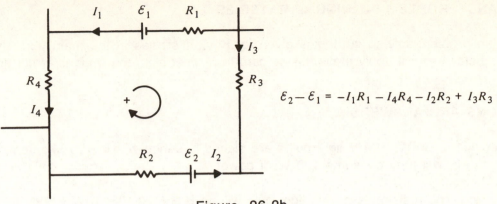

$$\mathcal{E}_2 - \mathcal{E}_1 = -I_1 R_1 - I_4 R_4 - I_2 R_2 + I_3 R_3$$

Figure 26-2b

$\mathcal{E}_1 - \mathcal{E}_2 = I_1 R_1 - I_4 R_4 + I_2 R_2 + I_3 R_3$ (from Fig. 26-2a)

$\mathcal{E}_2 - \mathcal{E}_1 = -I_1 R_1 - I_4 R_4 - I_2 R_2 + I_3 R_3$ (from Fig. 26-2b)

To convert a galvanometer whose full scale current is I_c and whose resistance is R_c into an ammeter whose full scale current is I, put it in parallel with a shunt resistance

$$R_{sh} = R_c[I_c/(I - Ic)]$$

To convert the same galvanometer into a voltmeter whose full scale voltage is V_{ab}, put it in series with a resistance

$$R = (V_{ab}/I_c) - R_c$$

An R-C series circuit has a time varying charge q and voltage i. If the capacitor is charging,

$$q = VC(1 - e^{-t/RC}) = Q_f(1 - e^{-t/RC})$$

$$i = \frac{dq}{dt} = \frac{V}{R} e^{-t/RC} = I_0 e^{-t/RC}$$

where V is the charging voltage, Q_f the final charge, and I_0 the initial current.

If the capacitor is discharged through a resistance of magnitude R,

$$i = I_0 e^{-t/RC}$$

$$q = Q_0 e^{-t/RC}$$

where Q_0 is the initial charge and I_0 the initial current.

HINTS AND PROBLEM-SOLVING STRATEGIES

Review the problem-solving strategies given for Kirchoff's rules in the main text. In the examples below we will apply these, taking care to respect sign conventions of currents and emf's.

QUESTIONS AND ANSWERS

Question. If two 50 W, 120 V light bulbs are placed in series with a 120 V potential drop across the pair, will they consume 100 W of power?

Answer. No. The resistance of one light bulb will be R = (120 V)^2/50 W = 288 Ω. The series resistance of the pair will be the sum of their individual resistances or 288 Ω + 288 Ω = 576 Ω. Therefore the power consumed by the pair in series will be P = (120 V)^2/576 Ω = 25 W. Since the voltage drop across each bulb was reduced by a factor of two using the series circuit, the power was reduced by four.

Question. If the same light bulbs in the previous question are now placed in parallel, will they consume 100 W of power?

Answer. Yes. Now each has a potential drop of 120 V across bulb so each consumes 50 W of power.

EXAMPLES AND SOLUTIONS

Example 1

Find the equivalent (R_{equ}) of the network of resistors in Fig. 26-3a.

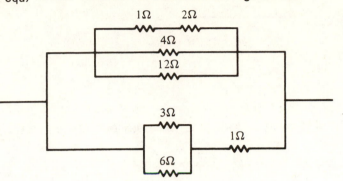

Solution:

Figure 26-3a

The successive reduction of the circuit is shown in Fig. 26-3 a,b,c. In Fig. 26-b the 1 Ω and 2 Ω in the upper branch of Fig. 26-3a have been combined according to the series rule,

$$1 \, \Omega + 2 \, \Omega = 3 \, \Omega,$$

whereas the 3 Ω and 6 Ω of the lower branch of Fig. 26-3a have been combined according to the rule for adding reciprocals,

$$\frac{1}{6 \, \Omega} + \frac{1}{3 \, \Omega} = \frac{1}{2 \, \Omega} \qquad R_{equ} = 2 \, \Omega$$

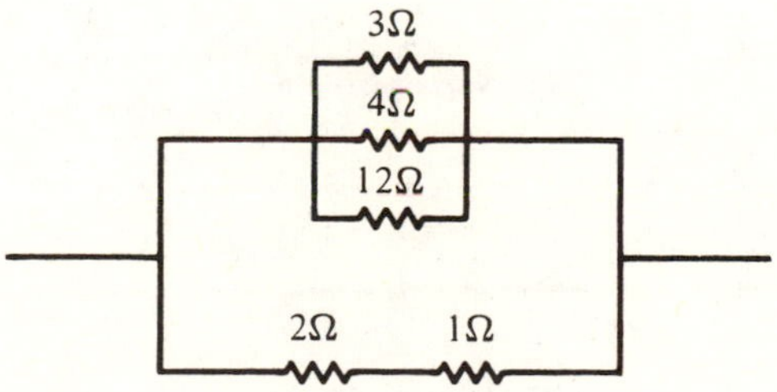

Figure 26-3b

In Fig. 26-3c the parallel 3 Ω, 4 Ω, and 12 Ω resistances in the upper branch of Fig. 26-3b have been combined according to the parallel rule

$$\frac{1}{3 \, \Omega} + \frac{1}{4 \, \Omega} + \frac{1}{12 \, \Omega} = \frac{8}{12 \, \Omega} = \frac{2}{3 \, \Omega}$$

$$R_{equ} = (3/2) \, \Omega$$

87

while the series resistances of 2 Ω and 1 Ω have been combined into an equivalent 3 Ω resistance.

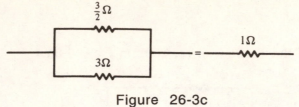

Figure 26-3c

Finally on the right side of Fig. 26-3c, the two parallel resistances have been combined,

$$\frac{1}{(3/2)\ \Omega} + \frac{1}{3\ \Omega} = \frac{1}{1\ \Omega} \qquad R_{equ} = 1\ \Omega$$

into the equivalent resistance 1 Ω.

Example 2

In the circuit of Fig. 26-4, V_{ab} = 12 V.
    (a) Find the currents I_1, I_2, and I_3.
    (b) Find the power dissipated in each resistance and in the entire network.

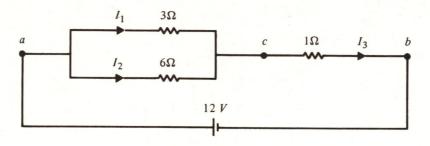

Figure 26-4

Solution:

According to Kirchhoff's rules,

$$I_1 + I_2 - I_3 = 0.$$

The parallel resistances can be reduced to 2 Ω (because 1/3 Ω + 1/6 Ω = 1/2 Ω), and this resistance can be combined with the 1 Ω resistance, resulting in an equivalent resistance (R_{equ}) of 2 Ω + 1 Ω = 3 Ω through which the current I_3 flows. Then

88

$$V_{ab} = 12 \text{ V} = R_{equ} I_3 = 3 \text{ }\Omega \text{ } I_3$$

$$I_3 = (12 \text{ V}/3 \text{ }\Omega) = 4 \text{ A}.$$

The potential drop across cb is thus

$$V_{cb} = I_3(1 \text{ }\Omega) = 4 \text{ V}.$$

Thus the potential drop V_{ac} is given by

$$V_{ac} + V_{cb} = 12 \text{ V}$$

$$V_{ac} = 12 \text{ V} - 4 \text{ V} = 8 \text{ V}.$$

Thus

$$V_{ac} = 8 \text{ V} = I_1(3 \text{ }\Omega); \qquad I_1 = 2.67 \text{ A}$$

$$V_{ac} = 8 \text{ V} = I_2(6 \text{ }\Omega); \qquad I_2 = 1.33 \text{ A}$$

Note the junction rule $I_1 + I_2 - I_3 = 0$ is obeyed.

The power dissipated in the entire network is

$$P = V_{ab}I_3 = I_3^2 R_{equ} \qquad P = V_{ab}I_3 = I_3^2 \text{ } R_{equ}$$

$$= 12 \text{ V}(4 \text{ A}) = 48 \text{ W}$$

The power dissipated in the 3 Ω, 6 Ω, and 1 Ω resistances is, respectively,

$$P_3 = I_1^2(3 \text{ }\Omega) = (2.67 \text{ A})^2(3 \text{ }\Omega) = 21.4 \text{ W}$$

$$P_6 = I_2^2(6 \text{ }\Omega) = (1.33 \text{ A})^2(6 \text{ }\Omega) = 10.6 \text{ W}$$

$$P_1 = I_3^2(1 \text{ }\Omega) = (4 \text{ A})^2(1 \text{ }\Omega) = 16.0 \text{ W}$$

Note $P_3 + P_6 + P_1 = 48$ W; the power dissipated in the entire network is equal to the sum of the power dissipated in each resistance.

Example 3

In the network of Fig. 26-5a,

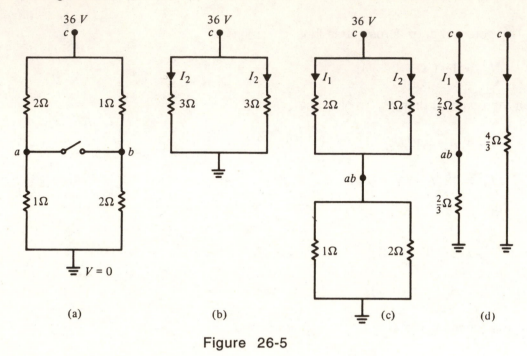

Figure 26-5

(a) Find the current in each branch when the switch is open.
(b) Find the potential drops V_{ca} and V_{cb} when the switch is open.
(c) Find the current in each branch when the switch is closed.
(d) Find the potential drops V_{ca} and V_{cb} when the switch is closed.

Solution:

Here is an exercise in thinking in terms of equivalent circuits.

(a) When the switch is open the circuit may be reduced to that of Fig. 26-5b. Thus

$$V_c = 36 \text{ V} = I_1(3 \text{ } \Omega) = I_2(3 \text{ } \Omega)$$

$$I_1 = I_2 = \frac{36 \text{ V}}{3 \text{ } \Omega} = 12 \text{ A}$$

(b) $$V_{ca} = I_1(2 \text{ } \Omega) = 12 \text{ A}(2 \text{ } \Omega) = 24 \text{ V}$$

$$V_{cb} = I_1(1 \text{ } \Omega) = 12 \text{ A}(1 \text{ } \Omega) = 12 \text{ V}$$

Thus

$$V_{ba} = V_{ca} - V_{cb} = 12 \text{ V}$$

(c) When the switch is closed (now $V_{ba} = 0$) the circuit may be successively reduced to the equivalent circuits in Fig. 26-5 c and d. Thus

$$V_c = 36 \text{ V} = I(4/3 \ \Omega)$$

$$I = \left(\frac{3}{4 \ \Omega}\right)(36 \text{ V}) = 27 \text{ A}$$

$$V_{ca} = I\left(\frac{2}{3} \ \Omega\right) = (27 \text{ A})\left(\frac{2}{3} \ \Omega\right) = 18 \text{ V} = V_{cb}$$

$$I_1 = \frac{V_{ca}}{2 \ \Omega} = \frac{18 \text{ V}}{2 \ \Omega} = 9 \text{ A}$$

$$I_2 = \frac{V_{ca}}{1 \ \Omega} = \frac{18 \text{ V}}{1 \ \Omega} = 18 \text{ A} \qquad \text{note:} \quad I_1 + I_2 = I$$

Example 4

In the circuit shown in Fig. 26-6, find the currents I_1, I_2, and I_3.

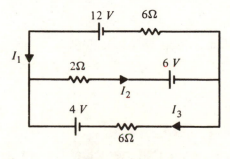

Figure 26-6

Solution:

Adopting the positive circuit conventions of Fig. 26-6, we write Kirchhoff's rules for each circuit loop, $\Sigma \mathcal{E} = \Sigma \ IR$:

(i) $12 \text{ V} - 6 \text{ V} = I_1(6 \ \Omega) + I_2(2 \ \Omega)$

(ii) $6 \text{ V} - 4 \text{ V} = -I_2(2 \ \Omega) - I_3 \ (6 \ \Omega)$

(iii) $12 \text{ V} - 4 \text{ V} = I_1(6 \ \Omega) - I_3(6 \ \Omega)$

91

These are three equations in three unknowns:

$$6 = 6I_1 + 2I_2$$
$$2 = -2I_2 - 6I_3$$
$$8 = 6I_1 - 6I_3$$

Are they independent? No, because the sum of the first two yields the third; an independent relation among I_1, I_2, and I_3 is given by Kirchhoff's junction rule

$$I_1 - I_2 + I_3 = 0$$

From the first two loop relations, we find

$$I_1 = \frac{6 - 2I_2}{6} \qquad\qquad I_3 = \frac{-2 - 2I_2}{6}$$

Substituting these in the junction rule yields

$$\frac{6 - 2I_2}{6} - I_2 + \frac{-2 - 2I_2}{6} = 0$$

$$I_2 = (2/5) \text{ A.}$$

Substituting this result in the first 2 loop relations yields

$$I_1 = (13/15) \text{ A}$$

$$I_3 = -(7/15) \text{ A.}$$

The fact that I_3 turns out negative means that the actual current direction is opposite to the arrow in Fig. 26-6.

To check these results, substitute them back into the original Kirchhoff's rules and see if they are verified.

Example 5

For the circuit of Fig. 26-7
   (a) Write down Kirchhoff's rules for the three subcircuits (i), (ii) and (iii) in terms of I_1, I_2, and I_3, as shown. Adopt the sign conventions as indicated by the circulating arrows.
   (b) Solve for I_1, I_2, and I_3.

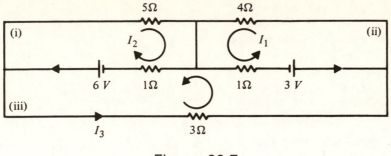

Figure 26-7

Solution:

(a) By Kirchhoff's junction rule, the current in the 5 Ω resistance is $I_2 - I_3$ and the current in the 4 Ω resistance is $I_1 + I_3$. Thus we have

(i) $6 \text{ V} = I_2(1 \ \Omega) + (I_2 - I_3)5 \ \Omega$

(ii) $3 \text{ V} = I_1(1 \ \Omega) + (I_1 + I_3)4 \ \Omega$

(iii) $(6 - 3)\text{V} = I_2(1 \ \Omega) - I_1(1 \ \Omega) + I_3(3 \ \Omega)$

(b) $I_1 = 0.15$ A $I_2 = 1.47$ A $I_3 = 0.56$ A

Note that this <u>cannot</u> be solved by parallel-series reduction methods.

Example 6

A voltmeter of resistance 500 Ω, placed in the circuit of Fig. 26-8 as shown, reads 50 V.
   (a) Find the resistance R and the currents I, I_1, and I_2.
   (b) What is the potential at point A when the voltmeter is not in the circuit?
(Equivalently, what does the voltmeter read if its resistance is infinite?)

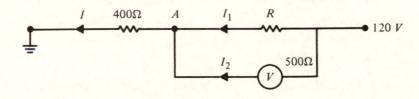

Figure 26-8

93

Solution:

(a) Point A is at a potential 120 V - 50 V = 70 V with respect to the ground where V = 0. Thus the current I through the 400 Ω resistor is given by

$$I(400 \ \Omega) = 70 \ V$$

$$I = (7/40) \ A$$

From Kirchhoff's junction rule we have

$$I_1 + I_2 = I = (7/40) \ A.$$

Across the parallel branches the potential drops are

$$I_1 R = 50 \ V$$
$$I_2(500 \ \Omega) = 50 \ V$$

The last three equations involve the unknowns R, I_1, and I_2. Their solution is

$$R = 667 \ \Omega \qquad I_1 = 0.075 \ A \qquad I_2 = 0.10 \ A$$

(b) The equivalent resistance of the two resistors (without the voltmeter) is

$$R_{equ} = 400 \ \Omega + 667 \ \Omega = 1067 \ \Omega.$$

The current through the circuit is

$$I = \frac{V}{R_{equ}} = \frac{120 \ V}{1067 \ \Omega} = 0.113 \ A$$

The voltage difference between the 120 V point and A is

$$V = RI = (667 \ \Omega)(0.113 \ A) = 75 \ V$$

(The voltmeter, having relatively low resistance, disturbs the circuit considerably. A better voltmeter will have a much higher resistance and read much closer to 75 V).

Example 7

The resistance of a galvanometer is 25 Ω. A current of 100 μA produces full scale deflection. Find the shunt resistance required to convert it to an ammeter reading 10 A at full scale.

Solution:

Referring to Fig. 26-9,

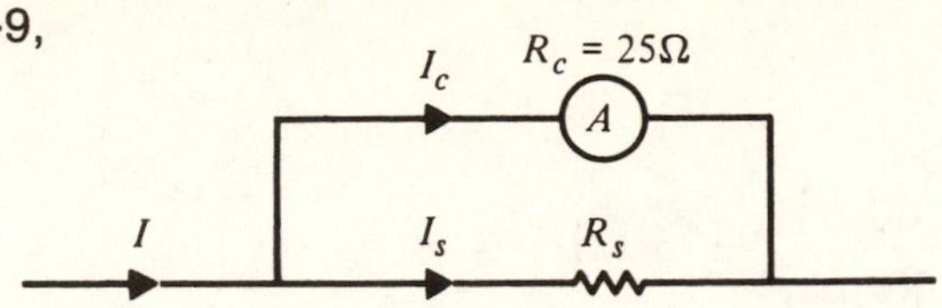

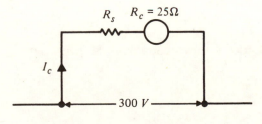

Figure 26-9

we wish to adjust the shunt resistance R_s so that

$$I_c = 100\ \mu A \text{ when } I = 10\ A. \text{ From } I_cR_c = I_sR_s \qquad \text{(parallel voltage drops equal)}$$

and $I_c + I_s = I$ (Kirchhoff's junction rule)

we see

$$R_s = R_c\left(\frac{I_c}{I_s}\right) = R_c\left(\frac{I_c}{I - I_c}\right)$$

$$R_s = \left(25\ \Omega\right)\left(\frac{100\ \mu A}{10\ A - 100\ \mu A}\right) \cong 2.5 \times 10^{-4}\ \Omega\ .$$

Example 8

Show how to convert the galvanometer of Example 7 to a voltmeter reading 300 V at full scale.

Solution:

Referring to Fig. 26-10,

Figure 26-10

we seek a shunt resistance R_s such that a current $I_c = 100 \ \mu A$ passes through the coil when 300 V is put across the shunt plus voltmeter combinations:

$$(R_s + R_c)I_c = 300 \ V$$

$$R_s = \frac{(300 \ V)}{I_c} - R_c = \frac{(300 \ V)}{(100 \ \mu A)} - (25 \ \Omega)$$

$$= 3 \times 10^6 \ \Omega$$

Example 9

The circuit of Fig. 26-11 is used to charge the two capacitors.
    (a) Find the final charge on each capacitor and the potential across each after the switch has been closed for a long time.
    (b) Find the time it takes for the charges and potential differences to reach half their final values.

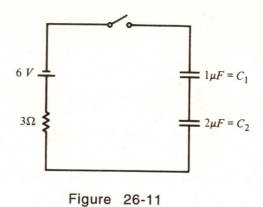

Figure 26-11

Solution:

(a) The equivalent capacitance of the two capacitors is given by

$$\frac{1}{C} = \frac{1}{1 \ \mu F} + \frac{1}{2 \ \mu F} = \frac{3}{2 \ \mu F}$$

$$C = \frac{2}{3} \ \mu F.$$

Thus the final charge on each is

$$Q_f = V_f C = (6 \text{ V})\left(\frac{2}{3} \times 10^{-6} \text{F}\right)$$

$$= 4 \times 10^{-6} \text{ C} = 4 \text{ } \mu\text{C}$$

The final potentials across the capacitors are

$$V_1 = \frac{Q_f}{C_1} = \frac{4 \text{ } \mu\text{C}}{1 \text{ } \mu\text{F}} = 4 \text{ V}$$

$$V_2 = \frac{Q_f}{C_2} = \frac{4 \text{ } \mu\text{C}}{2 \text{ } \mu\text{F}} = 2 \text{ V}$$

(b) The time constant of the circuit is

$$RC = 3 \text{ } \Omega \text{ } (4 \text{ } \mu\text{C}) = 12 \times 10^{-6} \text{ s.}$$

For a charging circuit, the charges and potentials across the capacitors vary as

$$Q = Q_f(1-e^{-t/RC})$$

$$V = \frac{Q}{C} = \left(\frac{Q_f}{C}\right)\left(1 - e^{-t/RC}\right) = V_f\left(1 - e^{-t/RC}\right)$$

For the charges and voltages to reach half their final values,

$$\frac{Q}{Q_f} = \frac{V}{V_f} = 1 - e^{-t/RC} = \frac{1}{2}$$

$$\frac{1}{2} = e^{-t/RC}$$

$$\ln(1/2) = - \ln 2 = -t/RC$$

$$t = RC \ln 2$$

$$= (12 \times 10^{-6} \text{ s})(0.69)$$

$$= 8.32 \times 10^{-6} \text{ s}$$

Example 10

You wish to install a radio in a car with a 12 V, negatively grounded electrical system. The power is to be drawn from the fuse box. Trace the complete circuit and the connections that need to be made.

Solution:

The negative terminal of the battery is connected to the ground, often the metallic chassis of the auto. Starting here we trace the circuit through the battery from ground (-) to hot (+) side to the fuse box where power may be distributed to any number of accessories, which are in parallel. A wire from the fuse box brings power to the radio's power supply on-off switch and through the radio to a ground wire, which completes the circuit to the negative side of the battery.

QUIZ

1. Three pieces of wire of the same length have resistances of 1 Ω, 2 Ω, and 3 Ω. They are braided together to produce a single resistor, which is connected across a 12 V battery. What is the current in each strand of the braided wire?

Answer: 12 A, 6 A, 4 A

2. Suppose that in Fig. 26-6a all emf's are kept the same, but each resistance is replaced by one with half its original resistance. What are the currents I_1, I_2, and I_3?

Answer: 13/30 A, 2/10 A, -7/30 A

3. The resistance of a galvanometer is 50 Ω. A current of 200 μA produces full scale deflection. Find the shunt resistance required to convert it to an ammeter reading 10 A at full scale.

Answer: 10^{-3} Ω

4. A 30 mF capacitor is charged by connecting it to a 12 V battery. The capacitor is then allowed to discharge by short circuiting its plates with a wire of resistance 20 Ω. How long does it take the charge on the capacitor to decay to 1/10 its original value?

Answer: 1.4 s

5. A galvanometer reaches full scale deflection when a current I_{fs} = 0.1 A passes through its coil of resistance R_c = 10 Ω. What shunt resistance is needed to convert it into a voltmeter with a maximum deflection of 20 V?

Answer: 190 Ω.

27
MAGNETIC FIELD
AND MAGNETIC FORCES

OBJECTIVES

In this chapter your objectives are to:

Calculate the vector force on a moving charged particle in a specified magnetic field using the right hand rule. The magnetic force does no work on charged particles.

Calculate the flux of a known magnetic field through a given surface.

Calculate the radius of the circular orbit that results when the particle velocity is perpendicular to the magnetic field.

Apply the expression for the magnetic force on a single charged particle to the calculation of the force on a conductor carrying current in a constant magnetic field.

Calculate the torque on a rectangular coil of wire carrying current in a constant magnetic field in terms of the magnetic moment of a current distribution.

Calculate the Hall emf (voltage) for a conductor in a constant, uniform magnetic field.

REVIEW

In this chapter the concept of a magnetic field is introduced in analogy to the electric field studied previously. A *static* charge produces an *electric field* which in turn produces a force on a second static charge. A *moving charge* or current produces a *magnetic field* which in turn causes a force to act on a second moving charge or current.

The electrostatic force depends on the particle's position and charge (q): Coulomb's law does *not* contain the particle's speed, $\vec{F}_E = q\vec{E}$. The magnetic force on the other hand depends not only on the speed, v, but also on the relative orientation of the velocity vector, $\vec{v}$, and the magnetic field, $\vec{B}$. This dependence is given by:

$$\vec{F}_M = q\vec{v} \times \vec{B}$$

If an electric field is also present ($\vec{E}$) then we must add these two forces vectorially:

$\vec{F}_{total} = q\vec{E} + q\vec{v} \times \vec{B}$.

The direction of the magnetic force is found by using the right hand rule to evaluate the cross product $\vec{v} \times \vec{B}$. Thus $\vec{F}_M$ is always perpendicular to both $\vec{v}$ and $\vec{B}$. One important consequence of the fact that F_M is perpendicular to $\vec{v}$ is that magnetic forces do no work on moving charged particles. Since $\vec{F}_M$ is perpendicular to $\vec{B}$, the lines of $\vec{B}$ are not lines of force as they were with the electric field but will be called *magnetic field lines*.

The infinitesimal flux, $d\Phi$, through a surface element dA is calculated by taking the product of the normal component of $\vec{B}$ (the tangential component does not contribute to the flux) with dA. For a closed surface, the normal ($\vec{n}$) is chosen to be the outward normal so that $B_n = \vec{B} \cdot \vec{n}$, where n is a unit vector. For an open surface, it will be defined more carefully in a later chapter. The total flux Φ, is obtained by integrating $d\Phi = \vec{B} \cdot \vec{n}$ dA over the surface in question. This is illustrated in Example 4.

The S.I. unit for magnetic field is the tesla. From the equation for the force, since qv has dimensions of coulomb·meters·(sec)$^{-1}$ and a coulomb per second is an ampere, one tesla is equal to one newton per ampere·meter. As flux is dimensionally a magnetic field multiplied by an area, it will have units of newton·meters per ampere. This unit is given the name of a weber. Hence one weber is equal to one tesla multiplied by one meter squared (1 T·m^2).

The motion of a charged particle in a constant magnetic field is very important to understand as it forms the basis for many later applications. The simplest way to understand this motion for an arbitrary orientation of the vectors $\vec{v}$ and $\vec{B}$ is to resolve v into components parallel to $\vec{B}$ (v_t) and perpendicular to $\vec{B}$ (v_n). No magnetic force is exerted because of v_t so that motion parallel to $\vec{B}$ proceeds at constant velocity. For the component $\vec{v}_n$, the magnetic force (of magnitude qv_nB) results in a uniform circular motion about the magnetic field direction. Since the motion associated with $\vec{v}_t$ is unaccelerated and that associated with $\vec{v}_n$ is uniform circular motion, the general trajectory is a helix (see Example 2). In the discussion to follow, we will omit the motion associated with $\vec{v}_t$ and concentrate just on the uniform circular motion. This will correspond to the special case where $\vec{v}$ is perpendicular to $\vec{B}$. We treat that case now.

In Example 3 it is shown that the magnetic force can do no work on a moving charged particle so the kinetic energy cannot increase or decrease (i.e. $\vec{v}$ is a vector of fixed length). The time rate of change of $\vec{v}$ or the acceleration **a** due to the magnetic force is thus perpendicular to $\vec{v}$ at each instant and points to the center of the circle. For uniform motion on a circle, the acceleration must have a magnitude of v^2/R where R is the radius. Since $\vec{v}$ is perpendicular to $\vec{B}$, the acceleration resulting from the magnetic force is:

$$a = \frac{qvB}{m}$$

For uniform circular motion, we must have:

$$\frac{qvB}{m} = \frac{v^2}{R}$$

Solving this for the radius we have: R = mv/qB. In our earlier study of uniform circular motion, we saw that there was an angular velocity, ω, associated with this motion such that Rω = v. For the charged particle in a uniform magnetic field, since ω = v/R, we deduce that ω = qB/m. This angular velocity or angular frequency is called the <u>cyclotron frequency</u>.

The period of revolution, τ, is equal to the distance around the circle divided by the speed, so τ = 2πR/v. Using the above value for R we find that τ = 2πm/qB or τ = 2π/ω. It is important to note that this period of revolution is independent of the particle's speed. Particles with higher speeds travel in larger circles but the time for a revolution is the same for all.

Applications of these ideas to the determination of the charge to mass ratio for electrons and the measurements of the relative masses of isotopes are excellent examples and discussed fully in the text.

The magnetic force on a single moving charge is used to calculate the *net* force on a conductor carrying current I in a uniform magnetic field B. Charge carriers of both signs (with different drift velocities) are considered and it is shown that the force always points in the same direction independent of the sign of the charge since the velocities are in opposite directions. The macroscopic current depends on the number of charge carriers per unit volume, the charge carried, the average drift velocity, and the cross-sectional area of the conductor. By adding up the individual forces on the charges, the total force is found to depend on B, I, and L, the length of conductor. If we define a vector $\vec{L}$ that points in the direction of the current flow and has magnitude equal to the length (L) considered, then:

$$\vec{F} = I(\vec{L} \times \vec{B}).$$

The use of this expression is illustrated in Examples 7b, 8, and 10.

If the direction (or magnitude) of the magnetic field varies over the length of the conductor, as it would if the conductor were not simply a long straight wire, then it is preferable to write the element of force, d$\vec{F}$, as,

$$d\vec{F} = I(d\vec{L} \times \vec{B})$$

where d$\vec{L}$ is a vector pointing in the direction of current flow and small enough so that B is essentially constant over its infinitesimal length, dL. This expression can then be integrated over the configuration of the conductor to obtain the total force. This approach is illustrated for a wire bent in the shape of a semi-circle in Example 9.

The magnetic force on a segment of conductor can be used to calculate the total force on a complete circuit in a constant, uniform magnetic field. This total force is zero and independent of the geometric shape of the circuit. See Example 7b. The net torque (Γ) however is not in general zero. This torque is first calculated just as it was in the study of mechanics by writing $\vec{\Gamma} = \vec{r} \times \vec{F}$ or more simply by finding the product of the force with the lever arm. This is illustrated in Example 7c. In the text this calculation is also done for a rectangular loop of wire. Once the torque has been calculated by finding $\vec{r} \times \vec{F}$, it can be recognized that by defining a new vector $\vec{M}$, the magnetic moment, the torque can be alternatively expressed by the cross product $\vec{M} \times \vec{B}$. Although the definition $\vec{M} = I\vec{A}$ was made with reference to the rectangular loop, it is perfectly general and applies to any shape of closed circuit. The method for proving this statement is indicated in the text and the proof itself is given in most advanced texts on electromagnetic theory. The potential energy of a magnetic moment in a magnetic field is a very useful expression for many calculations and it is given in the text.

The Hall effect is a consequence of the magnetic force on individual charge carriers. The internal non-electrostatic electric field set up by this force and the electrostatic force due to the charge separation necessary to maintain electrical neutrality is perpendicular to the magnetic field and the direction of current flow. The magnitude of this field is the same as that needed in a velocity filter (E = vB) except here the velocity is the average drift velocity. The direction of this field depends on the *sign* of the charge carrier. This is emphasized in the text and in Example 5. Measurement of the sign of the Hall emf can determine whether the majority charge carriers in a material are electrons or holes. Measurement of the magnitude of the Hall emf can determine the number of charge carriers per unit volume.

In addition to its usefulness in obtaining fundamental information about conductors and semiconductors, the Hall effect has some very practical applications. Since voltages are easy to measure with accuracy and the Hall emf is proportional to the magnetic field, "Hall probes" are used to measure laboratory magnetic fields. For the same reason, the Hall effect can be used to regulate magnetic fields by using the difference between the actual Hall emf and the desired Hall emf as an "error signal" to feed back to the current generator used to produce the magnetic field.

The number of extremely important practical devices that draw on the concepts presented in this chapter is very large. Three simple examples (the pivoted coil galvanometer, the d.c. motor, and the electromagnetic pump) were selected from this long list to underscore the practical nature of this material.

HINTS AND PROBLEM-SOLVING STRATEGIES

As in the study of mechanics, choice of a convenient coordinate system is essential.

1. Review the definition of the vector cross product and practice on Example 1.

2. Review the definition of the unit normal vector to an element of area dA.

3. Treat "closed surfaces", usually, as a collection of open surfaces and calculate B_n for each.

QUESTIONS AND ANSWERS

Question. Prior to the discovery in 1820 by Oersted that a current, initially parallel to a compass needle, deflected that needle, he used the same apparatus to demonstrate that electricity had no effect on magnetism. What was the original configuration of the wire and the compass needle?

Answer. The original orientation of the compass needle was <u>perpendicular</u> to the wire and hence parallel to the field produced by the wire so there was no effect when current was passed through the wire. When the compass needle is <u>parallel</u> to the wire before current passes through the wire, the deflection is most easily seen.

Question. Is the magnetic force fundamentally different from the electrical one given by Coulomb's law?

Answer. No. The two are manifestations of the same force, the electromagnetic force. The magnetic force can be thought of (and derived) as a relativistic correction to the Coulomb force.

EXAMPLES AND SOLUTIONS

Since nearly all the problems encountered here involve the magnetic force, which is proportional to the cross product of two vectors, it is essential to review the cross product discussion in Chapter 1.

Example 1

Given vectors $\vec{A} = 3\,\hat{\imath} + 4\,\hat{\jmath}$ and $\vec{B} = 2\,\hat{\jmath} + 3\,\hat{k}$,
     (a) calculate the vector product $\vec{A} \times \vec{B}$,
     (b) calculate the angle θ between $\vec{A}$ and $\vec{B}$,
Refer to Fig. 27-1.

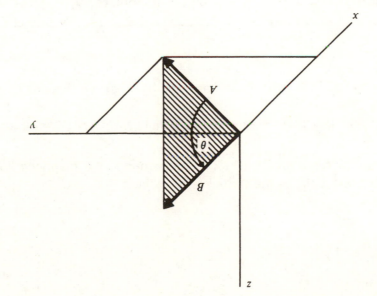

Figure 27–1

Solution:

(a) $\vec{A} \times \vec{B} = (3\,\hat{\imath} + 4\,\hat{\jmath}) \times (2\,\hat{\jmath} + 3\,\hat{k})$

          $= 3\,\hat{\imath} \times (2\,\hat{\jmath} + 3\,\hat{k}) + 4\,\hat{\jmath} \times (2\,\hat{\jmath} + 3\,\hat{k})$

103

$$= 6\hat{k} - 9\hat{\jmath} + 12\hat{\imath} \quad \text{(since } \hat{\imath} \times \hat{\jmath} = \hat{k}; \hat{\jmath} \times \hat{k} = \hat{\imath}; \text{ and } \hat{k} \times \hat{\imath} = \hat{\jmath})$$

$$= 12\hat{\imath} - 9\hat{\jmath} + 6\hat{k}$$

(b) To find θ we can use the fact that $\vec{A} \cdot \vec{B} = AB \cos\theta$ or that

$$|\vec{A} \times \vec{B}| = AB \sin\theta.$$

Using the first approach $(\vec{A} \cdot \vec{B} = AB \cos\theta)$, we note

$$\vec{A} \cdot \vec{B} = (3 \cdot 0 + 4 \cdot 2 + 0 \cdot 3) = 8$$

$$A = (3^2 + 4^2)^{1/2} = 5$$

$$B = (2^2 + 3^2)^{1/2} = (13)^{1/2}$$

which we combine in order to solve for the cosine

$$\cos\theta = (\vec{A} \cdot \vec{B})/AB = [8/5(13)^{1/2}] \quad \text{or } \theta = 63.7°.$$

The second approach gives

$$\sin\theta = [(261)^{1/2}/5(13)^{1/2}] \text{ or } \theta = 63.7°, \text{ as before.}$$

Example 2

A particle with charge $q = 1.6 \times 10^{-19}$ C, mass 2×10^{-27} kg, and velocity
$$\vec{v} = v_0(\hat{\imath} + 2\hat{\jmath} + 2\hat{k})$$
enters a region of constant and uniform magnetic field $\vec{B} = B_0\hat{\jmath}$. Given that $v_0 = 2 \times 10^5$ m/s and $B_0 = 2.5$ T,

  (a) Calculate the force on this moving charge due to the magnetic field.
  (b) Calculate the path of the particle in this field.

Solution:

(a) The force is given by

$$\vec{F} = q(\vec{v} \times \vec{B})$$

$$= qv_0B_0[(\hat{\imath} + 2\hat{\jmath} + 2\hat{k}) \times (\hat{\jmath})] \quad \text{(after substituting the given quantities)}$$

$$= qv_0B_0[\hat{k} + 2(-\hat{\imath})]$$

By squaring the components F_x and F_z and adding the squares, we find the magnitude of this force is equal to

$$|\vec{F}| = qv_0B_0(1 + 4)^{1/2} = qv_0B_0(5)^{1/2}$$

$$= qv_nB_0 .$$

where $v_n = (5)^{1/2}v_0$ is the magnitude of the velocity component perpendicular (or normal) to $\vec{B}$. Substituting the given values, the force is

$$|\vec{F}| = (1.6 \times 10^{-19} \text{ C})(2.5 \text{ T})(2 \times 10^5 \text{ m/s})(5)^{1/2}$$

$$|\vec{F}| = 1.79 \times 10^{-13} \text{ N}.$$

(b) The velocity component parallel to the field $v_0(2\mathbf{j})$ is *unaccelerated* so that the motion in the y direction proceeds at constant velocity.

In the xz plane the orbit is a circle with radius R obtained from

$$(mv_n^2/R) = qv_nB_0$$

$$R = \frac{mv_n}{qB_0} = \frac{(2 \times 10^{-27} \text{ kg})(\sqrt{5} \times 2 \times 10^5 \text{ m/s})}{(1.6 \times 10^{-19} \text{ C})(2.5 \text{ T})}$$

$$= 2.24 \times 10^{-3} \text{ m}.$$

The two motions when combined, the unaccelerated motion in the y direction and the circular motion in the xz plane, lead to a path called a helix.

Example 3

Calculate the work done on a charge q moving with velocity $\vec{v}$ in a constant magnetic field $\vec{B}$.

Solution:

The time rate of doing work (or the power) of a force, F, is given by the expression

$$P = (dW/dt) = \vec{F} \cdot \vec{v}$$

If we use for $\vec{F}$ the magnetic force, $\vec{F}_M = q\vec{v} \times \vec{B}$, then

$$\frac{dW}{dt} = q (\vec{v} \times \vec{B}) \cdot \vec{v} = 0$$

The right hand side is <u>zero</u> since $\vec{v} \times \vec{B}$ is a vector perpendicular to both $\vec{B}$ and $\vec{v}$ so its scalar product with $\vec{v}$ vanishes. Since dW/dt = 0, *no work is done by the magnetic force on charge q.* The work done on a charge q for a particular motion in an electric field was used to compute the change in potential energy. The above result for the magnetic field implies that there is no scalar quantity analogous to potential energy in electrostatics that characterizes the magnetic field.

Example 4

Calculate the flux through the surfaces ABCD and AEFD shown in Fig. 27-2 for a constant and uniform magnetic field $\vec{B} = 0.8$ T$(\hat{j})$.

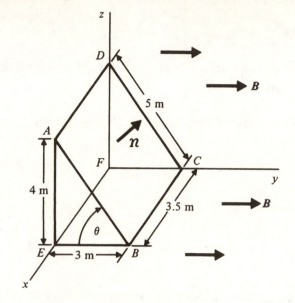

Figure 27-2

Solution:

The outward normal to the surface ABCD can be written:

$$\hat{n}_1 = \cos(90 - \theta)\hat{j} + \sin(90 - \theta)\hat{k}$$

$$= 0.8\,\hat{j} + 0.6\mathbf{k}$$

$$\vec{B} \cdot \hat{n}_1 = (0.8 \text{ T } \hat{j}) \cdot (0.8\,\hat{j} + 0.6\,\hat{k}) = 0.64 \text{ T}.$$

The flux through ABCD is

$$\Phi = (\vec{B} \cdot \hat{n}_1) A_{ABCD} = (0.64 \text{ T})(5 \text{ m})(3.5 \text{ m})$$

$$= 11.2 \text{ T·m}^2 = 11.2 \text{ webers}$$

The outward normal to the surface AEFD is equal to
$$\hat{n}_2 = -\hat{j}$$

106

The flux through this surface is

$$\Phi = (\vec{B} \cdot \hat{n}_2)A_{AEFD} = (0.8 \text{ T } \hat{\jmath})(-\hat{\jmath})(4 \text{ m})(3.5 \text{ m})$$

$$\Phi = -11.2 \text{ T·m}^2$$

In all, five surfaces constitute a closed surface in Fig. 27-2. The flux through *any* of the other surfaces (such as AEB, DFC, and EBCF) in Fig. 27-2 is *zero* because the normals to all the other surfaces are perpendicular to $\vec{B}$. Adding the two contributions calculated above, it is seen that the *total* flux through this closed surface is zero.

Example 5

A charge q moving with velocity v, perpendicular to a uniform magnetic field B is deflected from its original trajectory (where it would have hit the screen at S_0) and strikes the screen at S_1. What magnitude and direction of electric field E must exist between the plates P_1 and P_2 in order to give the charge a deflection of zero (i.e. to return it to S_0) while producing no <u>additional</u> forces? (This means that the <u>net force</u> acting on the charge <u>is zero</u>.) Refer to Fig. 27-3.

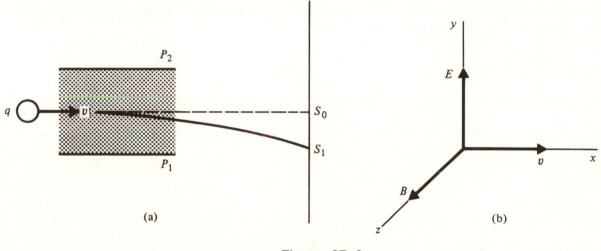

Figure 27–3

Solution:

Choose a coordinate system in which the initial velocity $\vec{v}$ points in the $\hat{\imath}$ direction, $\vec{v} = v\,\hat{\imath}$; the magnetic field points in the $\hat{k}$ direction, $\vec{B} = B\hat{k}$, and (if P_1 is positive with respect to P_2) $\vec{E} = E_x\,\hat{\imath} + E_y\hat{\jmath} + E_z\hat{k}$. The total force on the charge q from the combined electric and magnetic field (we ignore gravitation here) is:

$$\vec{F} = q\vec{E} + q\vec{v} \times \vec{B} = q[E_y\hat{\jmath} + vB(\hat{\imath} \times \hat{k})] + q[E_x\,\hat{\imath} + E_z\hat{k}]$$

$$= q[(E_y - vB)\hat{\jmath}] + q[E_x\,\hat{\imath} + E_z\hat{k}].$$

For zero deflection, this force must vanish, so $E_y = vB$; and $E_x = 0$, $E_z = 0$ are the conditions for $F = 0$. Therefore, an electric field pointing in the y direction and equal to the product vB is required.

Note that: (i) We have here a method for measuring the charged particle's velocity--by just measuring E and B since $v = E/B$; and (ii) We can use this arrangement of crossed (perpendicular) electric and magnetic fields as a *velocity-filter*.

If instead of a screen we had a small slit at S_0, only those particles whose velocity satisfied $v = E/B$ would pass through the slit. All others would suffer some deflection and be stopped.

Example 6

A given mass spectrometer can detect only ions that have a radius of curvature equal to that of helium of atomic mass 4, $_2\text{He}^4$, in a uniform magnetic field in which charged ions move in circles of radius R. How can the spectrometer be modified to make it detect helium of mass 3, $_2\text{He}^3$, assuming the atoms are singly ionized?

Solution:

Assume the spectrometer can detect only ions with a charge to mass ratio the same as a singly ionized helium atom of mass four because *the radius of curvature is fixed* by the instrument geometry and the fixed strength B_0 of the magnetic field. For circular motion in the constant magnetic field B_0 we have (with v perpendicular to B):

$$\frac{mv^2}{R} = qvB_0 \qquad \text{or} \qquad R = \frac{mv}{qB_0}$$

The only variable parameter in the expression for R is the velocity v in this example. (The velocity v is determined by the "filter" condition $v = E/B$ where B is a second constant magnetic field (not equal to B_0). Thus if m is reduced from four mass units to three mass units, v must be increased by a factor of 4/3 to keep R constant (i.e. $m_3v_3 = m_4v_4$). Thus $v_3 = (4/3)v_4$. This is best done by increasing the electric field E in the filter. For parallel plates $Ed = V$ where V is the potential difference and d is the plate separation. Thus the potential difference in the filter section should be increased by the same factor 4/3 to convert the spectrometer from mass 4 to mass 3.

Example 7

For the square coil geometry shown in Fig. 27-4,
- (a) calculate the flux through the open surface with the normal defined as shown;
- (b) calculate the force on segments ab, bc, cd, and da and show that F = 0 where F is the vector sum of these four forces;
- (c) calculate the torque about an axis parallel to the x axis through the center of the coil.

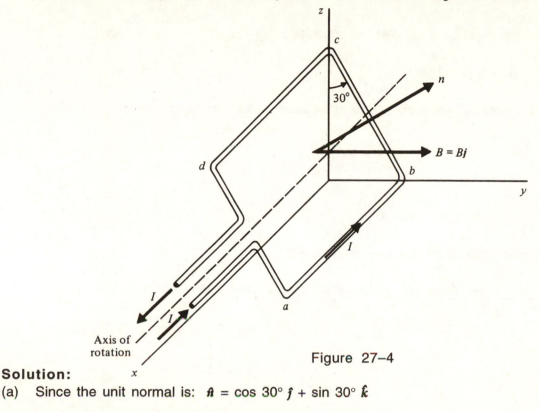

Figure 27–4

Solution:

(a)   Since the unit normal is: $\hat{n} = \cos 30°\, \hat{j} + \sin 30°\, \hat{k}$

$$\Phi = (\vec{B} \cdot \hat{n})A, \qquad \text{since B is constant.}$$

$$= |\vec{B}| \cdot |\hat{n}| \cdot \cos 30° = B \cos 30°$$

Thus since $A = L^2$ we have

$$\Phi = BL^2(\cos 30°)$$

(b) For segment ab, we have $\vec{L}_{ab} = L(-\hat{\imath})$ so

$$\vec{F}_{ab} = I\vec{L}_{ab} \times \vec{B} = IBL(-\hat{\imath} \times \hat{j}) = B\,IL(-\hat{k})$$

For segment bc,

$$\vec{L}_{bc} = L \sin 30° \,(-\hat{j}) + L \cos 30° \,(\hat{k})$$

109

so that the cross-product becomes

$$\vec{L}_{bc} \times \vec{B} = LB \cos 30°\ (\hat{\imath})$$

This leads to a force given by:

$$\vec{F}_{bc} = B\ IL\ \cos 30°\ (\hat{\imath})$$

For segment cd, we have $\vec{L}_{cd} = L\ \hat{\imath}$ (just the opposite of L_{ab}) so

$$\vec{F}_{cd} = -\ \vec{F}_{ab} = B\ IL(\hat{k})$$

For segment da, ignore the slight break due to the bend in the wires and note that

$$\vec{L}_{da} = -\ \vec{L}_{bd}$$

so that

$$\vec{F}_{da} = -\ \vec{F}_{bc} = -B\ IL\ \cos 30°\ (\hat{\imath}).$$

The forces cancel in pairs and the net magnetic force vanishes.

(c) Taking an edge view of the coil, since torque equals force times lever arm, we have, referring to Fig. 27-5,

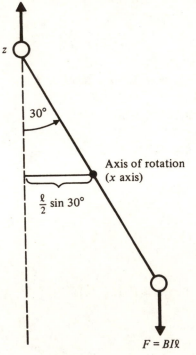

Figure 27–5

$$\Gamma = (B\ IL)\ [(L/2)\ \sin\ 30°]\ x\ 2$$

$$= B\ IL^2\ \sin\ 30°.$$

However

$$IL^2 = M,\ \text{the magnetic moment,}$$

$$\Gamma = BM\ \sin\ 30° = (1/2)B\ IL^2$$

where the torque has been taken about the x axis and has a contribution only from the forces on the dc and ab segments.

Another way to do this is to write the magnetic moment as a vector and calculate the torque $\vec{\Gamma}$,

$$\vec{M} = IA\hat{n} = IL^2\ [\cos\ 30°\ (\hat{j}) + \sin\ 30°\ (\hat{k})]$$

$$\vec{\Gamma} = \vec{M} \times \vec{B} = IL^2 B\ \sin\ 30°\ (\hat{k} \times \hat{j}) = B\ IL^2\ \sin\ 30°(-\hat{i}).$$

The vector $(-\hat{i})$ indicates that a clockwise rotation would take place about the x axis as counter-clockwise torques are positive.

Example 8

Find the force on a segment of straight wire, L, carrying a current, I, in a magnetic field B with arbitrary orientation with respect to the wire. Refer to Fig. 27-6.

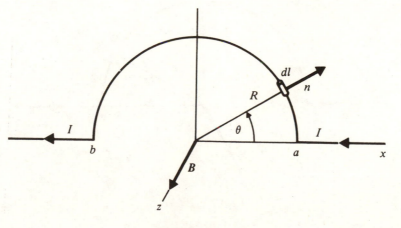

Figure 27–6

111

Solution:

Since B has arbitrary orientation, we write

$$\vec{B} = B_x\hat{\imath} + B_y\hat{\jmath} + B_z\hat{k},$$

and then by choosing different values for B_x, B_y, and B_z we can obtain any desired orientation of B. We have $\vec{L} = L(-\hat{\imath})$ since the current is in the negative x direction.

$$\vec{F} = I(\vec{L} \times \vec{B}) = IL \ (-\hat{\imath}) \times (B_x\hat{\imath} + B_y\hat{\jmath} + B_z\hat{k})$$

$$= IL \ [B_y(-\hat{k}) + B_z(\hat{\jmath})]$$

This is the final answer. We see that B_x is unimportant: the component of B parallel to the wire causes no force. In general then, the force vector lies in the YZ plane (the plane is perpendicular to the wire), the specific orientation depending on the relative magnitudes of B_y and B_z.

Example 9

Calculate the force on a wire of semi-circular shape carrying current I in a constant magnetic field B pointing in the + z direction. Refer to Fig. 27-7. Note that the force obtained is the same as that found for a straight wire connecting the end points a,b. See Example 8.

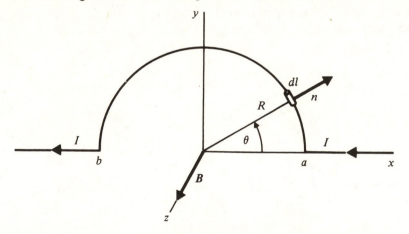

Figure 27–7

Solution:

The infinitesimal force on the segment dL carrying current I is

$$d\vec{F} = I\,d\vec{L} \times \vec{B}$$

Referring to Fig. 27-7, where $d\vec{L}$ is perpendicular to $\vec{B}$

$$d\vec{L} \times \vec{B} = (dL)(B)\hat{n}$$

where

$$\hat{n} = \cos\theta\,\hat{\imath} + \sin\theta\,\hat{\jmath}$$

Replacing dL by Rdθ, the element of force becomes

$$d\vec{F} = I(Rd\theta)(B)(\cos\theta\,\hat{\imath} + \sin\theta\,\hat{\jmath})$$

The total force on this semi-circle of wire is obtained by integrating $d\vec{F}$ from θ = 0 to θ = π.

$$\vec{F} = (BIR)\int_{0}^{\pi}(\hat{\imath}\cos\theta + \hat{\jmath}\sin\theta)\,d\theta$$

Note that

$$\int_{0}^{\pi}(\cos\theta)\,d\theta = 0$$

since sin θ vanishes at both limits, but that

$$\int_{0}^{\pi}(\sin\theta)\,d\theta = 2$$

Therefore after integration, we have

$$\vec{F} = 2B\,IR(\hat{\jmath})$$

Since we have 2R = diameter = length of wire between points a and b, this result is identical to Example 8.

If we consider a complete circle, then F = 0 because the sines and cosines will have both gone through a complete cycle.

Example 10

Suppose the magnetic field due to the earth can be represented by the vector

$$\vec{B} = B \cos 70° (-\hat{i}) + B \sin 70° (-\hat{k})$$

where $B = 2.5 \times 10^{-5}$ Tesla. A straight wire 1 m long carrying 500 A runs parallel to the ground (lies in the xy plane). Refer to Fig. 27-8.
     (a) Calculate the force when the current is in the (- x) direction.
     (b) Calculate the force when the current is in the (+ y) direction.

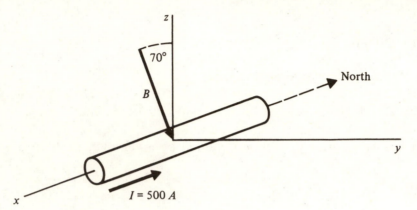

Figure 27–8

Solution:

(a) When the direction of the current is the - x direction,

$$\vec{L} = L(-\hat{i})$$

so that

$$\vec{F} = I\vec{L} \times \vec{B}$$

becomes

$$\vec{F} = IL(-\hat{i}) \times [B \cos 70°(-\hat{i}) + B \sin 70°(-\hat{k})]$$

$$= ILB \sin 70°(-\hat{j})$$

This force is in the negative y direction and of magnitude

$$|\vec{F}| = (500 \text{ A})(1 \text{ m})(2.5 \times 10^{-5} \text{ T})(0.940)$$

$$= 1.18 \times 10^{-2} \text{ N}.$$

114

(b) When the current is in the + y direction,

$$\vec{L} = L(\hat{\jmath})$$

The force becomes

$$F = IL(\hat{\jmath}) \times [B \cos 70°(-\hat{\imath}) + B \sin 70°(-\hat{k})]$$

$$= BIL[\cos 70°(\hat{k}) + \sin 70°(-\hat{\imath})]$$

$$= BIL[(0.940)(-\hat{\imath}) + (0.342)(\hat{k})]$$

making an angle of 20° with the (-x) direction in the xz plane. The magnitude of F is just BIL which equals 2.5 x 10⁻² N.

Since a transmission line carries an alternating current, the force will oscillate, causing the line to vibrate.

___

Example 11

A loop of wire carries current, I, to produce a magnetic moment, M, as in Fig. 27-9. The loop has mass m and its center of gravity is located L units from the z axis (the coil is pivoted in a frictionless manner about the z axis). A constant magnetic field $\vec{B} = B(-\hat{\imath})$ points in the (-x) direction. If θ is 37° what value of B is needed for equilibrium? Would this value of B give equilibrium for any value of θ?

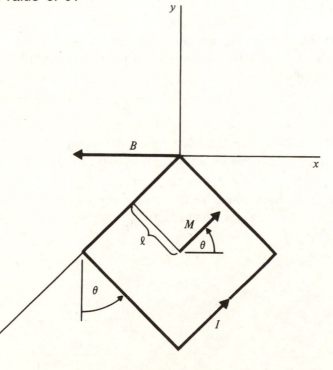

Figure 27–9

115

Solution:

$$\vec{B} = B(-\hat{\imath}) \text{ and } \vec{M} = M(\cos\theta\,\hat{\imath} + \sin\theta\,\hat{\jmath})$$

Therefore the torque on $\vec{M}$ due to $\vec{B}$ is:

$$\vec{\Gamma} = \mathbf{M} \times \vec{B} = BM\,(\sin\theta\,\hat{k}).$$

This tends to produce a c.c.w. rotation about the z axis. The torque due to the weight at the center of gravity, mg, is equal to $\mathbf{r} \times \mathbf{F}$ where

$$\vec{r} = L[\sin\theta\,(\hat{\imath}) + \cos\theta\,(-\hat{\jmath})]$$

is the vector to the center of gravity from the axis of rotation and $\vec{F} = mg(-\hat{\jmath})$. This torque is equal to:

$$\vec{\Gamma} = L[\sin\theta\,(\hat{\imath}) + \cos\theta\,(-\hat{\jmath})] = mgL\,\sin\theta\,(-\hat{k})$$

This would produce a c.w. rotation about the z axis. Requiring that

$$\Gamma_{total} = 0$$

we have:

$$BM\,\sin\theta - mgL\,\sin\theta = 0$$

Thus $\sin\theta$ cancels out, and if $B = (mgL/M)$, we have zero torque (and hence neutral equilibrium) at any angle! This result could have been anticipated by looking at the expressions for the potential energy.

Example 12

A fictitious material with 4×10^{28} charge carriers per m^3 (shown in Fig. 27-10) is in a constant magnetic field in the (-z) direction of 3 T. Let $x_1 = 3$ cm, $y_1 = 2.5$ cm, and $z_1 = 1$ mm.

    (a) If $I = 10$ A and the charge of each carrier is $q = -1.6 \times 10^{-19}$ C, calculate the magnitude and *sign* of the potential difference between the top face and the bottom face.

    (b) Do the same calculation but for positive charge carriers $q = +1.6 \times 10^{-19}$ C.

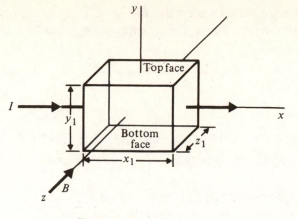

Figure 27-10

Solution:

To carry a current in the positive x direction, a negative charge carrier must have a drift velocity in the (- x) direction. $\vec{v} \times \vec{B}$ points in the (- y) direction but since the charge is negative, the magnetic force points in the (+ y) direction. Thus negative charge accumulates on the top face (and positive charge on the bottom face) until the internal electric field satisfies the equation E = vB. In this case, E points in (+ y) direction. The potential difference $V_H = Ey_1$ = vBy_1. The positive terminal of the voltmeter should be attached to the bottom face. To obtain the magnitude of V_H, eliminate v, the drift velocity by noting that the current per unit area (J) is equal to the product of nqv. Thus

$$v = \frac{J}{nq} = \frac{(I/y_1 z_1)}{nq}$$

and

$$V_H = vBy_1 = \frac{(IB)}{nqz_1}$$

Note V_H is independent of x_1 and y_1, but making the material thinner increases V_H!

Numerically, the product BI is 30 in S.I. units. Since force F = BIL, the units are N/m. We have

$$nqz_1 = (4 \times 10^{28} \text{ m}^{-3})(1.6 \times 10^{-19} \text{ C})(10^{-3} \text{ m}) = 6.4 \times 10^6 \text{ C/m}^2.$$

Thus

$$V_H = \frac{(30 \text{ N/m})}{\left(6.4 \times 10^6 \text{ C/m}^2\right)} = 4.69 \times 10^{-6} \text{ volts.}$$

(b) If the charge carriers were positive, the magnitude of V_H would remain the same but the sign would be reversed (top face is positive) since the magnetic force still points in (+ y) direction. Thus positive charge builds up on the top face. By measuring the sign of the Hall voltage V_H, one can determine whether the dominant charge carriers in a material are positive or negative.

QUIZ

1. A particle with charge $q = +3.2 \times 10^{-19}$ C, mass $m = 3 \times 10^{-27}$ kg, and velocity $\vec{v} = v_0\hat{j}$ where $v_0 = 5 \times 10^5$ m/s enters a region of constant and uniform magnetic field $\vec{B} = B_0\hat{i}$ where $B_0 = 1.6$ T.
    (a) Compute the magnitude of the magnetic force.
    (b) Calculate the radius of the resulting circular orbit.

Answer: (a) $F = 2.56 \times 10^{-13}$ N; (b) $R = 2.93 \times 10^{-3}$ m.

2. A negatively charged particle with velocity $\vec{v} = 3 \times 10^6$ m/s($\hat{i}$) enters a region of constant and uniform electric field $\vec{E} = E_0\hat{j}$ with $E_0 = 300$ N/C and constant and uniform magnetic field perpendicular to both $\vec{v}$ and $\vec{E}$. Calculate the magnitude and direction of $\vec{B}$ necessary to give the particle zero deflection.

Answer: $\vec{B} = (10^{-4}$ T$)(-\hat{k})$.

3. A square coil (side = 4 cm) with 12 turns lies in the xy plane. Its sides are parallel to the axes and its center is at the origin. A constant and uniform magnetic field exists in the + y direction with

$$\vec{B} = (2.0 \text{ T})\hat{j}.$$

Calculate the current needed to produce a net torque about the x axis of 1.92×10^{-3} N·m.

Answer: $I = 0.05$ A (remember there are 12 turns).

4. When there is a current of 1.5 A in a thin film (thickness = 10^{-6} m) a Hall voltage of 0.144 mv is developed across the face of width 1.2 cm. A 3.0 T magnetic field is applied perpendicular to the face.
    (a) Calculate the drift velocity of the charge carriers.
    (b) Calculate the current density, J.
    (c) Calculate the charge density of the carriers, nq.

Answer: (a) $v = 4 \times 10^{-5}$ m/s; (b) $J = 1.25 \times 10^8$ A/m^2; (c) $nq = 3.12 \times 10^{12}$ C/m^3.

5. A charge q moves in a region of space where B and E are constant and given by: $\vec{B} = (3$ T$)\hat{k}$ and $\vec{E} = (200$ N/C$)\hat{i}$. What velocity $\vec{v}$ is required for zero acceleration of the charge?

Answer: $\vec{v} = (- 66.7$ m/s$)\hat{j}$.

28
SOURCES OF MAGNETIC FIELD

OBJECTIVES

In this chapter your objectives are to:

Calculate the magnetic field at an arbitrary position in space produced by a single moving charge.

Calculate the contribution to the magnetic field at an arbitrary position in space due to a small element of current carrying conductor.

Apply Ampere's law to highly symmetric current distributions in order to calculate the magnetic field produced by that current distribution.

Distinguish between paramagnetism, diamagnetism, and ferromagnetism and learn how to modify the expressions obtained for the magnetic field due to a given current distribution if magnetic materials are present.

Define a *displacement current* and find its value in a charging or discharging capacitor and in a resistance carrying a varying current.

Specific problems that will be encountered include calculation of the magnetic field due to a long straight wire, the field due to a circular loop of current, the field due to a toroid, and the field at the center of a solenoid.

The cross products used in the last two chapters are used frequently here. The ideas of symmetry used in connection with Gauss's law are also very useful in the applications that involve Ampere's law.

REVIEW

In this chapter, three expressions, all equivalent, are presented for calculating the magnetic field due to a particular motion of the charges. The first expression, based on experimental observations is:

$$\vec{B} = \frac{\mu_0}{4\pi} q \frac{(\vec{v} \times \hat{r})}{r^2}$$

where $(\mu_0/4\pi) = 10^{-7}$ in S.I. units, $\vec{v}$ is the velocity of charge q and $\hat{r}$ is a unit vector from the position of the charge (source point) to the point where $\vec{B}$ is evaluated (field point).

Using the model previously employed in Chapter 28 to relate the force on a moving charge to the force on a conductor carrying current in a uniform magnetic field, the above expression is generalized to:

$$d\vec{B} = \frac{\mu_0}{4\pi} I \frac{(d\vec{l} \times \hat{r})}{r^2}$$

where as before $d\vec{l}$ is a vector pointing in the direction of current flow and of magnitude dl. This form is extremely useful since, when integrated, it gives the magnetic field at any desired point for any given current distribution. Example 2 illustrates this equation, particularly its vector nature. In the text this formula is applied to the current distributions presented by a long straight wire and a circular loop of current. These are two very important results as judicious application of these results to other, more complicated geometries, frequently leads to a good qualitative estimate of the magnetic field (even if an exact calculation isn't feasible).

For the long straight wire (assumed infinitely long) the lines of B form closed circles concentric with the wire. On any given circle of radius,

$$B = \left(\frac{\mu_0}{4\pi}\right)\frac{2I}{r}$$

An application of this formula with superposition of the fields of two currents is given in Example 1. In Example 4 the field due to a straight wire of finite length is calculated and this result is shown to agree with the above formula as the length becomes infinite. For the circular loop, the magnetic field is calculated in the text only on a line perpendicular to the plane of the coil and passing through its center. On this line, B is parallel to the line (but not so, off the line). The result for B on this line is found to be

$$B = \left(\frac{\mu_0}{4\pi}\right)\frac{2I(\pi a^2)}{(a^2 + x^2)^{3/2}}$$

where a is the radius of the loop and $(a^2+x^2)^{1/2}$ is the distance from the field point to any point on the loop. Since $I(\pi a^2)$ is the magnetic moment M, of the current loop,

$$B = \left(\frac{\mu_0}{4\pi}\right)\frac{2M}{(a^2 + x^2)^{3/2}}$$

This form gives a very good indication of how quickly B falls off with increasing distance from the loop. Again it is seen that the lines of B (although not circles in this case) are closed curves with no beginning and no ending. This is always true for magnetic field lines and must be contrasted with the electric field lines which begin on positive charges and end on negative ones. For the magnetic field, there are no point sources, so the lines of B are closed. As there are no "sources" or "sinks" of the magnetic field, *the flux of B through any closed surface vanishes.*

Since the magnetic field lines for a long straight wire are concentric circles about the wire, the force on a second wire carrying current and parallel to the first is easy to calculate. This is illustrated in Example 3. In this case, B produced by the first wire is perpendicular to the current in the second wire so the force on the second wire is either attractive or repulsive depending on the relative current directions. If the two wires carry current in the *same* direction, they *attract* whereas if the currents are in *opposite* directions, they repel.

The magnitude of the force on wire 2 carrying current I_2 is:

$$F_2 = B_1 I_2 L_2$$

where B_1 is the magnetic field due to wire 1 evaluated at the position of wire 2. If r is the separation between the two parallel wires, then

$$B_1 = \left(\frac{\mu_0}{4\pi}\right)\frac{2I_1}{r}$$

and

$$F_2 = \left(\frac{\mu_0}{4\pi}\right)\frac{2I_1 I_2 L_2}{r}$$

The force per unit length of wire 2 has the symmetric form,

$$\frac{F_2}{L_2} = \left(\frac{\mu_0}{4\pi}\right)\frac{2I_1 I_2}{r}$$

This expression provides the practical definition of the ampere in the S.I. system. If both I_1 and I_2 are equal to 1 ampere and r is 1 meter, then the force per unit length on either conductor is 2×10^{-7} N. The Coulomb is then defined as an ampere-second (i.e. the quantity of charge that flows past a point in one second if the current is 1 ampere).

The third method of calculating the magnetic field is based on *Ampere's Law*, a relationship analogous to Gauss's law for electrostatics. In the text, this law is justified by appealing to the resultant B for a long straight wire.

$$B_{wire} = \left(\frac{\mu_0}{4\pi}\right)\frac{2I}{r}$$

The lines of B point in a direction tangent to the circle of radius r. Thus

$$\oint \vec{B} \cdot d\vec{r} = B \cdot 2\pi r = (\mu_0/4\pi)\, I.$$

or

$$\oint \vec{B} \cdot d\vec{r} = \frac{\mu_0}{4\pi}\, I.$$

121

As stated above, this result is totally general and applies to any geometry. In the above equation, I is the current that passes through the surface contained in the contour around which the line integral was performed. Just as we could have started electrostatics by writing down Gauss's law and deriving everything else (including Coulomb's law) from it, we could have started this chapter by writing Ampere's law and obtaining all the other results from it.

As elegant and general as Ampere's law is, its usefulness in calculating B for a given current distribution is confined to cases where the symmetry is so high that the direction and/or magnitude of B is constant on the various parts of the contour so that no actual integration has to be performed. If you cannot exploit the symmetry of the current distribution to write down the value of the line integral $\oint \vec{B} \cdot d\vec{r}$ from physical grounds, then Ampere's law is not useful in obtaining B. Three examples are considered in the text (in addition to the long straight wire): the solenoid, the toroid, and the field between a charging parallel plate capacitor. Examples 5, 6 and 7 deal with aspects of Ampere's law.

A *displacement current density* exists everywhere there is a changing electric field; its direction is the direction of the change of E and its magnitude is

$$J_D = \varepsilon_0 \left(\frac{dE}{dt} \right).$$

Between the plates of a parallel plate capacitor the *displacement current* written as

$$I_D = J_D A = \varepsilon_0 \left(\frac{dE}{dt} \right) A = \varepsilon_0 \frac{d}{dt} \left(\frac{\sigma}{\varepsilon_0} \right) A$$

takes the specific value

$$I_D = \frac{d}{dt} (\sigma A) = \frac{dQ}{dt} = I_C$$

where I_c is the conduction current leading into the capacitor. Thus the total current, which is the sum of the displacement current and the conduction current, is continuous, the same across the capacitor gap as in the connecting wires.

Real materials can be classified qualitatively into three magnetic types: ferromagnetic materials, which are strongly attracted to the magnetic field produced by an electromagnet (or permanent magnet); paramagnetic materials, which are weakly attracted to the same magnetic field; and diamagnetic materials, which are weakly repelled by the above magnetic field.

This classification can be made more quantitative by forming the material in question into a "Rowland ring" and using a toroidal wrapping of wire on this ring. All the flux produced is confined by this geometry and the ratio of the magnetic field produced inside the material (ring), B, to the field that would be produced in vacuum, B_0, by the same current and number of turns per unit length is defined to be the relative permeability, K_m. Thus $K_m = B/B_0$. K_m is exactly equal to one for a non-magnetic material, slightly greater than one for a paramagnetic material, slightly less than one for a diamagnetic material, and much larger than one for a ferromagnetic material. The field inside the material is related to the current distribution which produced it by the equation:

$$B = K_m \mu_0 (N/L)I = \mu(N/L)I,$$

where $\mu = K_m\mu_0$.

HINTS AND PROBLEM-SOLVING STRATEGIES

(1) Because vector cross-products are used frequently in this chapter, choice of a convenient coordinate system is essential. (2) Check the symmetry of the problem in order to see which method of calculating B is applicable. (3) Substitute numerically, only at the last step. (4) Review the problem-solving strategies of the main text.

QUESTIONS AND ANSWERS

Question. Can you use Ampere's law to find the magnetic field at the center of a ring of radius R carrying current I?

Answer. No. A path where $\vec{B}$ is parallel to $d\vec{r}$ and has the same magnitude all along the path, can't be found.

Question. Is there a magnetic force between two identical charges with identical parallel velocities?

Answer. Yes. It is an attractive force, just as it is for two identical, parallel, current elements.

EXAMPLES AND SOLUTIONS

Example 1

Two long straight wires as shown in the two sketches of Fig. 28-1a and 28-1b carry currents of 100 A and 200 A in the indicated directions. What is the magnetic field at point P midway between the wires if they are 0.1 m apart?

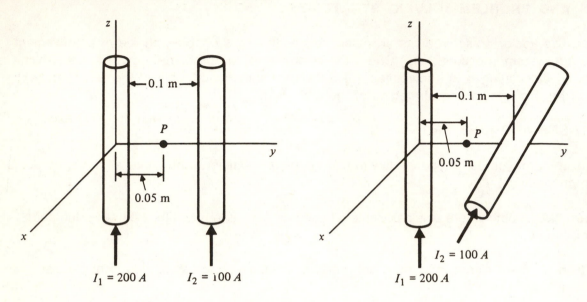

Figure 28-1

Solution:

(a) The magnetic field due to currents 1 and 2 should be calculated separately and then added (superimposed). The field at P due to current 1 points in the (- x) direction (as can be seen from the right hand rule) and is of magnitude:

$$B_1(P) = \left(\frac{\mu_0}{4\pi}\right)\frac{2I_1}{r}$$

where $(\mu_0/4\pi) = 10^{-7}$ T·m/A, $I_1 = 200$ A and r = 0.05 m; numerically we have

$$\vec{B}_1(P) = -8 \times 10^{-4} \text{ T } (\hat{x})$$

The field at P due to current 2 points in the (+ x) direction and is half the magnitude of $B_1(P)$ because the current is half as great. Thus

$$\vec{B}(P) = -4 \times 10^{-4} \text{ (I)}$$

124

(b) The field at P due to current 1 is unchanged in magnitude and direction:

$$\vec{B}_1(P) = -8 \times 10^{-4} \text{ T } (\hat{\imath})$$

The field due to current 2 has the same magnitude at P as in part (a) but the <u>direction is different</u>, superimposing (adding) these results,

$$\vec{B}_2(P) = 4 \times 10^{-4} \text{ T}(\hat{k}).$$

Thus we have

$$\vec{B}(P) = 4 \times 10^{-4} \text{ T}(\hat{k}) - 8 \times 10^{-4} \text{ T}(\hat{\imath})$$

This is a vector of magnitude 8.94×10^{-4} T making an angle of 26.6° with the (− x) axis.

Example 2

Use the Biot law to calculate the magnetic field at point P due to the segments of current carrying conductors shown in Fig. 28-2. Take the current as I and the radius of the circle as R.

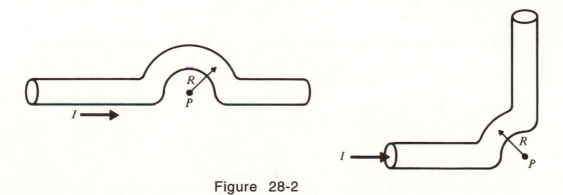

Figure 28-2

Solution:

Consider the segment to the left in Fig. 28-2. We divide that conductor into three regions, two straight lines and a semi-circle of radius R. The final answer is obtained by superposition using the problem-solving strategy of the main text. The vector relationships in the Biot law simplify these calculations.

Straight-line segments. Since P is on the line of the straight segments in both configurations of Fig. 28-2, the vector cross-product $d\vec{L} \times \hat{r} = 0$ because the vectors $d\vec{L}$ and $\hat{r}$ are parallel or anti-parallel. Therefore, these straight-line segments make no contribution to the magnetic field at P.

Semi-circle. Due to the above result, only the portions of the circle contribute to the field at P. For the current direction chosen, $|d\vec{L} \times \hat{r}| = dL$ and the direction is into the paper. Thus

$$dB = \left(\frac{\mu_0}{4\pi} I\right) \frac{dL}{R^2}$$

but $dL = R \, d\theta$ and for the configuration to the left (part a) in Fig. 28-2, the integration over θ is from 0 to π whereas in the configuration to the right (part b) in Fig. 28-2, the integration covers 0 to $\pi/2$. Thus we have

$$B = \left(\frac{\mu_0}{4\pi}\right) \int \frac{IR \, d\theta}{R^2} = \left(\frac{\mu_0}{4\pi}\right)\left(\frac{I}{R}\right) \int d\theta$$

$$B = \left(\frac{\mu_0}{4\pi}\right)\left(\frac{I}{R}\right)(\theta_f - \theta_i)$$

The answers for parts (a) and (b) are:

(a) $\quad B(P) = \left(\frac{\mu_0}{4\pi}\right)\left(\frac{I\pi}{R}\right)$; $\qquad$ (b) $\quad B(P) = \left(\frac{\mu_0}{4\pi}\right)\left(\frac{I\pi}{2R}\right)$

Example 3

Calculate the force on the various parts of the rectangular circuit carrying current I_2 shown in Fig. 28-3 due to the magnetic field of the infinitely long straight wire carrying current I_1.

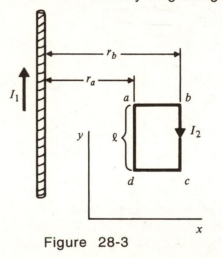

Solution: $\qquad\qquad\qquad\qquad$ Figure 28-3

(a) On segment ab, the field due to I_1 is into the paper and of magnitude

$$B_{ab} = \left(\frac{\mu_0}{4\pi}\right)\left(\frac{2I_1}{r}\right)$$

126

where r ranges from r_a to r_b. $d\vec{L} \times \vec{B}$ points in the (+ y) direction. Since B varies with position, we should write $dF = I_2 \, dr \, B(r)$ and integrate over r

$$\vec{F}_{ab} = \left(\frac{\mu_0}{4\pi}\right)\left(2\,I_1I_2\right)(\hat{\jmath})\int_a^b \frac{dr}{r}$$

$$\vec{F}_{ab} = \left(\frac{\mu_0}{4\pi}\right)\left(2\,I_1I_2\right)\left(\ln\frac{r_b}{r_a}\right)(\hat{\jmath})$$

(b) On segment bc, $d\vec{L} \times \vec{B}$ points in the (+ x) direction and B due to I_1 is constant in value and equal to

$$B_1 = \left(\frac{\mu_0}{4\pi}\right)\frac{2I_1}{r_b}$$

so the force is

$$\vec{F}_{bc} = B_1I_2L\,(\hat{\imath}) = \left(\frac{\mu_0}{4\pi}\right)\left(\frac{2\,I_1I_2L}{r_b}\right)(\hat{\imath})$$

(c) The considerations in part (a) apply here except now the order of integration is reversed so

$$\vec{F}_{cd} = -\,\vec{F}_{ab}$$

Thus there is no net force in the y direction.

(d) This is just like part (b) except

$$B_1 = \left(\frac{\mu_0}{4\pi}\right)\frac{2I_1}{r_a}$$

along path da. Thus we have

$$\vec{F}_{da} = \left(\frac{\mu_0}{4\pi}\right)\left(\frac{2\,I_1I_2L}{r_a}\right)(-\hat{\imath})$$

Since $r_a < r_b$, there is a net force in the (- x) direction:

$$\vec{F} = \vec{F}_{bc} + \vec{F}_{da}$$

$$\vec{F} = \left(\frac{\mu_0}{4\pi}\right)\left(2\,I_1I_2L\right)\left[\frac{1}{r_b} - \frac{1}{r_a}\right](\hat{\imath})$$

Example 4

Use the result for the magnetic field produced by a straight segment of wire (length L) at the midpoint to calculate the field at the center of a square coil. Refer to Fig. 28-4.

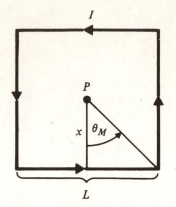

Figure 28-4

Solution:

The field at P due to one segment is equal to

$$B = \left(\frac{\mu_0}{4\pi}\right) \frac{2I}{x} \sin \theta_M$$

For a square coil as shown in Fig. 28-4, B points out of the page. The contributions from all four segments are equal and $\theta_m = 45°$. Thus by superposition:

$$B(P) = 4 \left(\frac{\mu_0}{4\pi}\right) \frac{I}{L/2} \frac{\sqrt{2}}{2} = 4\sqrt{2} \left(\frac{\mu_0}{4\pi}\right) \left(\frac{I}{L}\right).$$

This field is larger than the field at the center of a circular loop of diameter L by the factor $(2)^{1/2}$.

Example 5

Use the result for the magnetic field due to a long straight wire to evaluate the integral of $\vec{B} \cdot d\vec{r}$ round the closed contour (abcda) shown in Fig. 28-5 in the plane perpendicular to the wire.

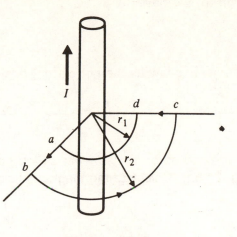

Figure 28-5

Solution:

(a) On segment ab, $\vec{B}$ is perpendicular to $d\vec{r}$ so $\vec{B} \cdot d\vec{r} = 0$.

(b) On segment bc, $\vec{B}$ is parallel to dr at every point so $\vec{B} \cdot d\vec{r} = B\, dr$. Furthermore at the distance r_2 from the wire we have

$$B = \left(\frac{\mu_0}{4\pi}\right)\frac{2I}{r_2}$$

which is the same magnitude at each point on segment bc. The value of B can be taken outside the integral leaving only the integral of dr over the quarter circle of radius r_2. This gives $(1/4)(2\pi r_2)$ for the path length, implying

$$\int \vec{B} \cdot d\vec{r} = \left(\frac{\mu_0}{4\pi}\right)\left(\frac{2I}{r_2}\right)\left(\frac{\pi r_2}{2}\right) = \left(\frac{\mu_0}{4\pi}\right)(\pi I)$$

(c) For the segment cd, once again $\vec{B}$ is perpendicular to $d\vec{r}$ so $\vec{B} \cdot d\vec{r} = 0$.

(d) On the segment da, $\vec{B}$ is antiparallel to dr so $\vec{B} \cdot d\vec{r} = -B\, dr$. B is constant in magnitude and equal to

$$B = \left(\frac{\mu_0}{4\pi}\right)\frac{2I}{r_1}.$$

The integral over dr gives $\pi r_1/2$ so:

$$\int_{da} \vec{B} \cdot d\vec{r} = -\int_{da} B \cdot dr = -\left(\frac{\mu_0}{4\pi}\right)\left(\frac{2I}{r_1}\right)\left(\frac{\pi r_1}{2}\right) = -\left(\frac{\mu_0}{4\pi}\right)(\pi I)$$

Thus if we add results (a) through (d) we have

$$\oint \vec{B} \cdot d\vec{r} = 0.$$

This is the result expected from Ampere's law as the <u>net current enclosed by the contour abcda is zero.</u>

Example 6

Equal and opposite currents, I, are carried on the surfaces of long concentric <u>thin</u> cylinders of radii R_1 and R_2, as shown in Fig. 28-6a. Calculate the magnetic field between the cylinders and outside the two cylinders.

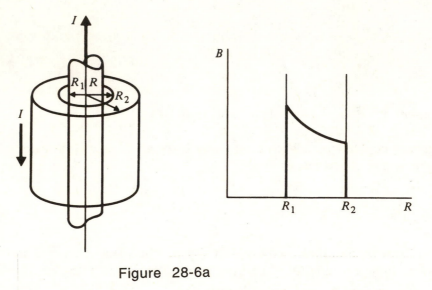

Figure 28-6a

Solution:

Choose a circular contour of radius R where $R_2 > R > R_1$. By symmetry, $\vec{B}$ is tangent to the circle of radius R and parallel to a c.c.w. path element $d\vec{L}$. Also $\vec{B} \cdot d\vec{r} = B \, dr$ and B is constant on this circle. Thus

$$\oint \vec{B} \cdot d\vec{L} = B * dr = B \cdot 2\pi R = \mu_0 I,$$

where I is the current carried by the inner conductor. Solving for B we obtain

$$B = \left(\frac{\mu_0}{4\pi}\right)\frac{2I}{R}$$

which is the same result that one would obtain for a long straight wire. This result is shown in the sketch given in Fig. 28-6A--which shows the field for values of R between R_1 and R_2.

For a circular contour with $R > R_2$, while the direction of B is less obvious since there are two currents penetrating the surface, it is clear from rotational symmetry that B has the same magnitude at each point on the circle. The direction of $\vec{B}$ must be along the tangent to the circle since the individual contributions from the two cylinders are each separately tangent to the circle. Designating this constant tangential component of $\vec{B}$ as B_t, leads to

$$B_t \cdot 2\pi R = \mu_0 \text{ (current enclosed)} = \mu_0(I - I) = 0.$$

Thus we have $B_t = 0$ outside the second conductor ($R > R_2$), similar to electric field calculations for this geometry from Gauss's law. The field is then zero for $R < R_1$ and $R > R_2$.

There is one additional result we could obtain from this problem. Suppose we remove the inner conductor, leaving only a hollow cylinder of radius R_2 carrying current, I. For the imaginary surface of radius $R < R_2$ (inside the shell), the same symmetry conditions apply as were discussed previously but now no current cuts this surface. So $B \cdot 2\pi R = 0$ now and $B = 0$.

This result would not hold if the current flowed around the cylindrical surface in circular loops as shown in Fig. 28-6b. Then there would be a magnetic field inside the cylinder pointing along the axis of the cylinder since this geometry is essentially the same as that of a solenoid.

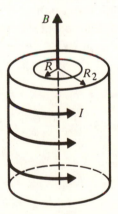

Figure 28-6b

131

Example 7

Use Ampere's law to calculate the magnetic field both inside and outside a solid cylindrical conductor of radius R that carries current I where the current per unit area (J) is constant. Refer to Fig. 28-7.

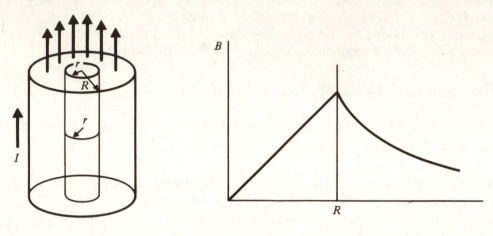

Figure 28-7

Solution:

Choose a circle of radius r < R. The region exterior to this circle gives no contribution to B on the circle. This was established in the previous example. For the region interior to the circle, B is tangent to the circle (the long straight wire result) and constant in magnitude. So

$$\oint \vec{B} \cdot d\vec{r} = \int B\, dr = B \int dr = B \cdot 2\pi r$$

The current enclosed by this path is equal to $J \cdot \pi r^2$ since the current per unit area is assumed constant. Thus we have

$$2\pi r B = \mu_0 J \pi r^2$$

Solving for B we find

$$B = \left(\frac{\mu_0}{4\pi}\right) J(2\pi r) \qquad\qquad r \leq R$$

This solution is valid up to the point r = R where we write: (note $I = \pi R^2 J$)

$$B(R) = \left(\frac{\mu_0}{4\pi}\right)\left(\frac{2J}{R}\right)(\pi R^2)$$

$$B(R) = \left(\frac{\mu_0}{4\pi}\right)\left(\frac{2I}{R}\right), \qquad \text{a familiar result.}$$

132

For a circle of radius $r > R$, the same symmetries apply and

$$2\pi r\, B = \mu_0 I$$

since this path encloses all of the current. In this region

$$B = \left(\frac{\mu_0}{4\pi}\right)\left(\frac{2I}{R}\right)$$

Thus B is zero at the center of this current distribution, increases linearly with r, the distance from the center, until $r = R$ (where B is maximum) and then decreases as r^{-1} for $r > R$. This is shown in the sketch in Fig. 28-7.

Example 8

A Rowland ring with circumference of 0.25 m is constructed from a material with relative permeability of 500. The winding consists of 1000 turns. For a current of 3 A, calculate
  (a) the magnetic intensity H;
  (b) the magnetic field B;
  (c) the magnetization.

Solution:

(a) The magnetic intensity $H = nI$ where n is the number of turns per unit length and I is the current.

$$H = (1000/0.25\ \text{m})(3\ \text{A}) = 1.2 \times 10^4\ \text{A/m}$$

(b) The magnetic field $B = \mu H = K_m\mu_0 H = 4\pi K_m(\mu_0/4\pi)H$ where $K_m = 500$; thus

$$B = 4\pi(500)(10^{-7}\ \text{N/A}^2)(1.2 \times 10^4\ \text{A/m}) = 7.54\ \text{T}.$$

(c) Since $B = \mu_0(H + M)$, $M = (B/\mu_0) - H$, and

$$M = \frac{\left(\frac{B}{4\pi}\right)}{\left(\frac{\mu_0}{4\pi}\right)} - H = \frac{(7.54\ \text{T})}{4\pi\left(10^{-7}\ \text{N/A}^2\right)} - \left(1.2 \times 10^4\ \text{A/m}\right)$$

$$= 5.99 \times 10^6\ \text{A/m}$$

Example 9

For the capacitors shown in Figure 28-8, calculate the displacement current in the capacitor gaps and show it is equal to the conduction current in the wire. (See Example 9 of Chapter 26).

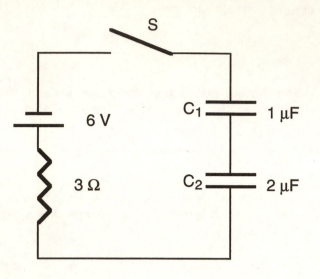

Figure 28-8

Solution:

The electric field in the capacitors is

$$E = \frac{\sigma}{\varepsilon_0} = \frac{Q}{A\varepsilon_0}$$

The displacement current is

$$I_D = J_D A = \varepsilon_0 \left(\frac{dE}{dt}\right) A .$$

Since

$$E = \frac{\sigma}{\varepsilon_0} = \frac{Q}{A\varepsilon_0} ; \qquad \frac{dE}{dt} = \frac{1}{A\varepsilon_0} \frac{dQ}{dt}$$

We have for I_D,

$$I_D = \varepsilon_0 \left(\frac{dQ}{dt}\right)\left(\frac{A}{A\varepsilon_0}\right) = \frac{dQ}{dt} = I_c$$

$$I_D = \frac{d}{dt} Q_f\left(1 - e^{-t/RC}\right)$$

$$I_D = \frac{Q_f}{RC}\left(e^{-t/RC}\right) = I_c$$

134

Where I_c is the conduction current in the wire (assumed perfectly conducting; see the next problem for I_D in a real conductor.)

Example 10

Find the displacement current in the 3 Ω copper resistor of Example 9 (above) at time t = 0.

Solution:

Across the resistor we have

$$V = IR = EL$$

where E is the field along the current flow and L the length of the current flow. Thus within the resistor,

$$E = \frac{IR}{L}$$

The displacement current is

$$I_D = J_D A = A\varepsilon_0\left(\frac{dE}{dt}\right) = A\varepsilon_0\left(\frac{R}{L}\right)\left(\frac{dI}{dt}\right)$$

$$I_D = \frac{AR}{L}\,\varepsilon_0\left(\frac{dI}{dt}\right) = \rho\varepsilon_0\left(\frac{dI}{dt}\right)$$

where $\rho = (AR/L)$ is the resistivity of the copper resistor. We have $\rho = 1.72 \times 10^{-8}$ Ω·m,

$$\frac{dI}{dt} = \frac{d}{dt}\left(\frac{Q_f}{RC}\,e^{-t/RC}\right) = -\frac{Q_f}{(RC)^2} \quad \text{at t = 0;}$$

$$I_D = -\rho\varepsilon_0\left(\frac{Q_f}{(RC)^2}\right)$$

Thus the displacement current in the wire is quite small:

$$I_D = -\left(1.72 \times 10^{-8}\ \Omega\cdot m\right)\left(8.85 \times 10^{-12}\ C^2/N\cdot m^2\right)\left(\frac{4 \times 10^{-6}\ C}{\left(12 \times 10^{-6}\ s\right)^2}\right)$$

$$= -4.2 \times 10^{-15}\ A = -4.2 \times 10^{-3}\ pA.$$

QUIZ

1. Two long straight wires carry currents I_1 = 150 A and I_2 = 200 A *in the same direction*. The wires are parallel and separated by d = 0.2 m.

  (a) Calculate the field at the position of wire 1 and the force per meter due to I_2;

  (b) Calculate the field at the position of wire 2 and the force per meter due to I_1;

  (c) Show that these two forces are equal in magnitude but opposite in direction. Do they have the same line of action?

Answer: (a) B = 2 x 10^{-4} T, F/L = 3 x 10^{-2} N/m; (b) B = 1.5 x 10^{-4} T, F/L = 3 x 10^{-2} N/m; (c) They are equal and opposite and have the same line of action.

2. For which of the configurations listed below can Ampere's Law be used to calculate the magnetic field?
(a) A wire of infinite length; (b) A wire of finite length; (c) A solenoid at the end; (d) A solenoid at the center; (e) A toroid; and (f) A circular loop of wire.

Answer: (a), (d), and (e).

3. A toroid consists of N turns wound uniformly on a "doughnut" shaped form. A constant current of I amps is maintained in the winding. If the inner radius is 23 cm and the outer radius is 25 cm, use Ampere's law to estimate the percentage deviation of the magnetic field from the value at the mean radius of 24 cm.

Answer: Around the circle of radius 23 cm, the field is 4.35% higher than at the center. Around the circle of radius 25 cm, the field is 4 % lower than at the center.

4. The relative permeability of the core material in a Rowland Ring is 5000. The *mean* circumference of the ring is 0.4 m. The winding consists of 2000 turns and carries a current of 2 A. Calculate

  (a) the magnetic intensity H

  (b) the magnetic induction B

  (c) the magnetization M

Answer: (a) H = 10^4 A/m

            (b) B = 62.8 T

            (c) M = 5.0 x 10^7 A/m

5. A charge q = - 1.6 x 10^{-19} C orbits an equal but opposite charge in a circle of radius r = 0.5 x 10^{-10} m at a constant speed of 2.25 x 10^6 m/s. (a) Calculate the magnitude of the current produced by the orbiting charge; and (b) the magnitude of the magnetic field produced at the center of the circle.

Answer: (a) I = 1.15 mA and (b) B = 14.4 T.

29
ELECTROMAGNETIC INDUCTION

OBJECTIVES

In this chapter, you are introduced to Faraday's law of electromagnetic induction. Your objectives are to:

Calculate the motional emf that arises from the magnetic force.

Calculate the magnetic flux through an open surface.

Calculate the time rate of change of flux due to a change in effective area or due to a change in the magnetic field itself (or both).

Determine the direction of current in a closed circuit when an induced emf is present.

Summarize the laws of electromagnetism in the form of Maxwell's equations.

REVIEW

In previous chapters, the magnetic force on a moving charge particle has been used to predict orbits in magnetic fields, forces and torques on current carrying conductors, the Hall effect, and a variety of other interesting physical phenomena. Here it is used to establish the existence of a motional electromotive force (emf) for a conductor moving in a magnetic field. The effect of the magnetic force is shown to be equivalent to a non-conservative electric field, E_n, of magnitude $E_n = vB \sin \theta$ where θ is the angle between the velocity v and the field B. If the conductor is not part of a complete circuit, there will be no current but an emf will exist in the conductor. This emf is defined in terms of a line integral of the scalar product of E_n with dl the element of path length.

$$\mathcal{E} = \int \vec{E}_n \cdot d\vec{l}$$

If the moving conductor forms part of a complete circuit (in which there are no other sources of emf) then there is a current, I. The magnitude of I is given by: $I = \mathcal{E}/R$. This current leads to a force F on the conductor that can be calculated using the methods in Ch. 28 and 29. For a straight wire of length L moving in a constant field, B, this force is $F = BIL$ and is in such a direction as to oppose the original motion. This last conclusion is reasonable as the opposite situation would have this force assisting the motion so that the velocity of the moving conductor would increase--which would further increase the force and we would have a "run-away" solution that would violate energy conservation. Experience tells us this is not the case so the force must oppose the motion. The above situation is analysed point-by-point in Example 1.

Another case of great practical importance is that of a coil of wire rotating about an axis perpendicular to a constant external magnetic field. An emf is induced in this case because the angle between the magnetic field and the normal to the plane of the coil changes with time. This time dependent change of flux ($\Phi = \vec{B} \cdot \hat{n}\, A$ in this case) results in an emf that can be taken off the rotating coil (depending on the type of ring used) as either an alternating voltage or a pulsating voltage all of one sign. This device then is a generator and converts mechanical energy (needed to rotate the coil) into electrical energy.

Faraday's law, namely

$$\mathcal{E} = -\frac{d\Phi}{dt}$$

is a generalization of the previous idea which states that an emf is developed if there is relative motion between a conductor and a magnetic field. No motion is necessary in Faraday's law, only a time dependent magnetic flux, in order to generate an emf. Thus if the flux through a stationary coil is changed by increasing or decreasing the magnetic field, an emf is generated. This type of emf (no moving parts) is illustrated in Examples 3, 4, and 5.

Several comments need to be made regarding Faraday's law. If the coil has N turns rather than one turn, then the flux through the entire coil

$$\Phi = N \int \vec{B} \cdot \hat{n}\, dA$$

and is N times larger than the flux through 1 turn (assuming all turns are alike). The emf is thus increased by a factor N. If Φ_1 is the flux through 1 turn only (and all turns are alike), then:

$$\mathcal{E} = -N\frac{d\Phi_1}{dt}$$

Since $\Phi = N\Phi_1$, the equation just written is identical in content to the previous one.

The non-conservative electric field, E_n, which is responsible for the emf in Faraday's law,

$$\mathcal{E} = \oint \vec{E}_n \cdot d\vec{L} = -\frac{d\Phi}{dt}$$

is very different from the conservative electric fields that arise from static charge distributions. For such conservative electric fields, the line integral (like the one above) of $\vec{E}_n \cdot d\vec{L}$ around a closed curve always gives zero. Furthermore the integral of $\vec{E} \cdot \hat{n}\, dS$ over a closed surface is equal to Q/ε_0 where Q is the charge inside the surface (Gauss's law). For our non-conservative electric field, E_n, the line integral around a closed curve is not necessarily zero but depends on the time rate of change of flux. The integral of $\vec{E}_n \cdot \hat{n}$ dS over any closed surface, however, is *always zero*.

In our previous discussion of induced emf's, we have been mainly concerned with the magnitude of this emf. Lenz's law is concerned with the sign of the emf. Lenz's law states that the direction (sign) of the induced emf is such as to oppose the motion (or change of flux) that created the emf. The sign convention associated with the calculation of the flux and the positive sense of the line integral (essentially a right hand rule) is carefully discussed in the text and should be adhered to in all problems.

The laws of electromagnetism are neatly summarized in a set of equations known as Maxwell's equations. We have met each of these equations before individually when we studied Gauss's law, the magnetic equivalent of Gauss's law, Ampere's law (with the displacement current included), and Faraday's law. These four equations, together with the Lorentz force law describe all of electromagnetic theory.

HINTS AND PROBLEM-SOLVING STRATEGIES

(1) Review the problem-solving strategy in the main text on Faraday's Law.

(2) Review the procedure for calculation of the magnetic flux through a given surface. Emphasis here is on the change of flux but first, in most problems, you must calculate the flux.

(3) To get the sign correct for the induced current, make an initial guess and then use the right-hand-rule for the current direction you assumed to find the direction of the B field resulting from the induced current. If the direction of this B field opposes the change in flux that induced the current, then your assumption was correct, otherwise, reverse the current direction. See Example 10.

QUESTIONS AND ANSWERS

Question. A copper ring that has been placed between the poles of a strong magnet so that its normal is parallel to the field requires force to remove it even though the copper is non-magnetic. Explain.

Answer. Due to eddy currents induced (due to flux changes) in the ring as the copper is withdrawn, there will be a retarding force as predicted by Lenz's law.

Question. The current in a long straight wire, along the x-axis of the x-y plane, changes sinusoidally with time, inducing a voltage in a rectangular coil lying in the x-y plane. How will the direction of the induced $\vec{E}$ be related to the direction of the magnetic field ($\vec{B}$) produced by the current?

Answer. $\vec{E}$ will be perpendicular to $\vec{B}$.

EXAMPLES AND SOLUTIONS

Example 1

If the conducting rod AB of length 0.25 m in Fig. 29-1 is moving to the right with a velocity of 3 m/s in a constant magnetic field of magnitude 0.8 T (pointing into the paper), what external force must be applied to maintain constant velocity? Assume the complete circuit has a total resistance R = 0.2 Ω.

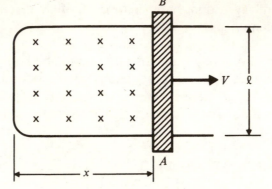

Figure 29-1

Solution:

The magnetic flux increases in this example because more area is being added to the region of constant magnetic field enclosed by the conducting loop with the rod at the right end.

(a) When the rod's distance from the left edge of the loop is x, the flux through the loop is:

$$\Phi = BA = B\,(Lx)$$

Using Faraday's law to calculate the emf,

$$\mathcal{E} = -\frac{d}{dt}\Phi = -\frac{d}{dt}(BLx) = -BLv$$

Numerically,

$$|\mathcal{E}| = BLv = (0.8\ T)(0.25\ m)(3\ m/s) = 0.6\ volts$$

Since the resistance is R, we obtain for the current I

$$I = \frac{BLv}{R} = \frac{\mathcal{E}}{R} = \frac{0.6\ V}{0.2\ \Omega} = 3\ A.$$

(b) This current flows from A to B producing a field for the motion shown so as to the decrease the change in flux. A magnetic force $\vec{F}_m = I\vec{L} \times \vec{B}$ is exerted on this conductor in a direction opposite to $\vec{v}$. Since $\vec{L}$ and $\vec{B}$ are perpendicular,

$$F_m = BIL = B\left(\frac{vBL}{R}\right)L = \frac{(BL)^2 v}{R} = 0.6 \text{ N}.$$

using the value of I obtained in (a). This force opposes the motion and points to the left.

(c) To move the rod AB to the right with constant velocity, v, we must supply an external force, to the right, with magnitude equal to F_m given above.

Example 2

Show that 1 volt is equal to 1 weber per second.

Solution:

The volt was defined as the work done per unit positive charge so one volt is equal to one joule per coulomb.

$$1 \text{ V} = \left(\frac{1 \text{ J}}{1 \text{ C}}\right)$$

The joule is a newton·meter whereas the coulomb is one ampere·sec. Thus

$$1 \text{ V} = \left(\frac{1 \text{ N·m}}{1 \text{ A·s}}\right)$$

The tesla is defined so that 1 T equals one newton per ampere·meter.

$$1 \text{ V} = \left(\frac{1 \text{ N·m}^2}{1 \text{ A·m·s}}\right) = \left(\frac{1 \text{ T·m}^2}{1 \text{ s}}\right) = \left(\frac{1 \text{ weber}}{1 \text{ s}}\right)$$

Since the weber is defined to be 1 T·m².

Example 3

A high power line carrying a current of $I(t) = I_0 \cos 2\pi ft$ is located near a rectangular coil with dimensions of 5 cm x 10 cm as shown in Fig. 29-2. I_0 = 250 A and f = 60 Hz. Calculate the emf induced in this coil.

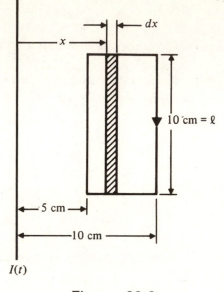

Figure 29-2

Solution:

To solve this problem, first we must use the result for the magnetic field near a long straight wire to calculate the flux through the coil. Then we use Faraday's law to obtain the emf from the time rate of change of the flux.

On the shaded element of area, $dA = L\,dx$, the field has the value

$$B = \frac{\mu_0}{4\pi} \frac{2I(t)}{x}.$$

For the direction chosen for I, B points inward so if we choose $\hat{n}$ inward, the flux at t = 0 is positive. This means that the positive sense of dL for the line integral is clockwise.

$$\Phi = \int B\,dA = \frac{\mu_0}{4\pi} 2I(t)L\int_{x_i}^{x_f} \frac{dx}{x}$$

$$\Phi = \frac{\mu_0}{4\pi} 2I(t)L \ \ln\left(\frac{x_f}{x_i}\right)$$

Since $x_f/x_i = 2$, we have

$$\Phi = \frac{\mu_0}{4\pi} \, 2I(t)L \, \ln(2)$$

Using Faraday's Law, we have

$$\mathcal{E} = \frac{d\Phi}{dt} = \frac{\mu_0}{4\pi} \, 2L \, \ln(2)(2\pi f) \, I_0 \, \sin 2\pi f t$$

At t = 0, this is positive, so the non-conservative electric field dotted into dL is positive on our contour and the initial induced current flows c.w.

It is instructive to calculate the peak value of this emf to get some idea of the problem posed by "pick-up" near power lines.

$$\mathcal{E}_{peak} = (10^{-7} \, T\cdot m/A)(1.386)(.1 \, m)(377 \, s^{-1})(250 \, A) = 1.31 \times 10^{-3} \, T\cdot m^2/s = 1.31 \, mV$$

If the coil had N turns (instead of one), this result would be increased by this factor N, i.e.

$$\mathcal{E}_{peak} = (1.31 \, N) \, mV \, .$$

Suppose the nearest edge of the coil is 1 m away from the power line rather than 5 cm. Then x_f/x_i = 1.05/1.00 and the factor ln 2 (= 0.693) is replaced by ln 1.05 (= 0.0488). This results in a reduction in peak emf by a factor of 7×10^{-2}.

Example 4

A square coil (l = 0.25 m) is placed around a solenoid of average diameter 0.1 m and length (L) of 0.2 m wrapped with 1000 turns (N) as shown in Fig. 29-3. The coil has its normal parallel to the axis of the solenoid. The current through the solenoid when it is energized obeys the equation

$$I(t) = I_0(1 - e^{-t/\tau})$$

where I_0 is equal to 100 A and τ is 5 seconds. Calculate the emf induced in the square coil as a function of time.

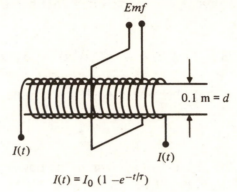

$$I(t) = I_0 (1 - e^{-t/\tau})$$

Figure 29-3

Solution:

(a) First we need the relationship between the current in the solenoid and the magnetic field B. Using the value of B appropriate to the center of a long solenoid,

$$B = \frac{\mu_0}{4\pi} \, 4\pi \left(\frac{N}{L}\right) I \ .$$

Thus the magnetic flux through the square coil is equal to the above value of B multiplied by the *area of the solenoid, $\pi d^2/4$,* not the area of the square coil. (We assume that the solenoid is long enough so that B is essentially zero outside the solenoid volume so that the only region of the square coil that contains magnetic field lines is the part inside the solenoid.)

(b) The emf is found by calculating $d\Phi/dt$.

$$\mathcal{E} = \frac{\mu_0}{4\pi} \, 4\pi \left(\frac{N}{L}\right) \frac{\pi d^2}{4} \left(\frac{dI}{dt}\right)$$

$$\mathcal{E} = \frac{\mu_0}{4\pi} \, 4\pi \left(\frac{N}{L}\right) \left(\frac{\pi d^2}{4}\right) \left(\frac{I_0}{\tau}\right) e^{-t/\tau}$$

The peak value of the emf is obtained at t = 0. This peak value is equal to 9.87 x 10⁻⁴ volts. From t = 0, the emf decreases exponentially with increasing time. If the square coil contains N turns then the emf will be increased by this factor of N.

Example 5

For the stationary single turn coil (r = 10 cm) shown in the Fig. 29-4 calculate (a) the flux through the coil, (b) the emf induced, and (c) the current in the coil if its resistance is R = 0.1 Ω. Let the time dependence of the magnetic field be given as $\vec{B} = B_0 \exp(-t/\tau)(\hat{\jmath})$ where $B_0 = 1.5$ T and $\tau = 3$ s.

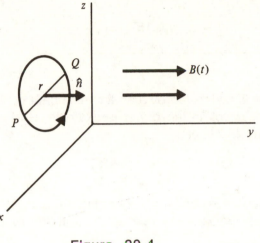

Figure 29-4

Solution:

If we choose the normal to the coil $\hat{n} = \hat{\jmath}$ (as shown), then the positive sense of $d\vec{L}$ is c.c.w. as you face the xz plane.

(a) The flux through the coil is:

$$\Phi = B(t)A = B(t)\,\pi r^2$$

(b) The emf is found from Faraday's law

$$\mathcal{E} = -\frac{d\Phi}{dt} = -(\pi r^2)\frac{dB}{dt} = -(\pi r^2)\left(-\frac{1}{\tau}B_0 e^{-t/\tau}\right)$$

Thus

$$\mathcal{E} = (\pi r^2)\left(\frac{B_0}{\tau}\,e^{-t/\tau}\right).$$

This emf has a peak value of $\pi r^2 B_0/\tau$ and decays to zero exponentially. Since $\mathcal{E}$ is positive (or zero), the line integral of $\vec{E}_n \cdot d\vec{l}$ is positive so $\vec{E}_n$ is parallel to $d\vec{l}$ on this circle. Thus the current direction is in the positive sense of dl or c.c.w. here. Again, the flux produced by the current tends to oppose the changing flux of the field. (Lenz's law)

(c) The current $I = \mathcal{E}/R$ so

$$I_{peak} = (\pi r^2)\frac{B_0}{R\tau}$$

For the numerical values given here,

$$\mathcal{E}_{peak} = 1.57 \times 10^{-2} \text{ V and } I_{peak} = 0.157 \text{ A.}$$

Example 6

Suppose an ideal voltmeter is connected by a straight line path between two points P and Q on opposite sides of the circular coil in the previous example. (Refer to Fig. 29-4). Calculate the potential difference read by this meter.

Solution:

For this coil, emf is generated uniformly throughout the circle so no specific part of the coil acts as the source of emf. Similarly the resistance is uniformly distributed around the ring. Thus writing

$$V = \mathcal{E} - IR$$

where V is the potential difference, $\mathcal{E}$ is the emf, and I the current, one sees that $V_{PQ} = 0$. This results from I being equal to $\mathcal{E}/R$. Note that on the path from P to Q, we develop half the total emf and encounter half of the total resistance so

$$V_{PQ} = (1/2)\mathcal{E} - I(1/2 \text{ } R) = 1/2(\mathcal{E} - IR) = 0$$

as $I = \mathcal{E}/R$.

Example 7

Suppose the ring in the previous example is made of two different metals, the resistance of one part being 0.1 Ω while the other resistance is 0.2 Ω so the total resistance is now 0.3 Ω. Refer to Fig. 29-5. The voltmeter connection is the same as in Example 6. If the induced emf is 0.06 V with the current flowing c.c.w., would the potential difference between P and Q read by the voltmeter still be zero?

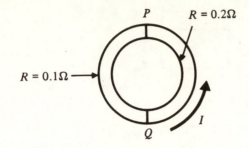

Figure 29-5

Solution:

The current I is continuous and equal to 0.2 A and the emf is uniformly generated but the losses are different on the two halves of the ring. Thus V_{PQ} = 0.03 V - (0.2)(0.2)V or V_{PQ} = 0.01 V with P negative with respect to Q.

Example 8

A coil with L = 12 cm and w = 6 cm and mass per unit length of 10⁻² kg/m is dropped from a height h as shown in Fig. 29-6. It enters the region of constant magnetic field of B = 0.5 T with the velocity acquired in the fall. The coil resistance is 0.1 Ω. What height h should be chosen so that the velocity of the coil in the magnetic field is constant until the entire coil is in the constant magnetic field?

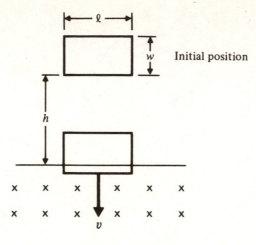

Figure 29-6

Solution:

(a) The flux through the coil, when the edge of the coil is y units into the field is BLy.

(b) The time rate of change of flux is BLv where v = dy/dt so this is the emf. The current is equal to the emf divided by the resistance.

$$I = BLv/R$$

(c) The force on this current carrying wire is upward and equal to

$$F_m = BIL = \frac{(BL)^2 v}{R}$$

(d) If the velocity is to remain constant, then this upward force must be just equal to the weight, mg. (i.e. the coil is in free fall until the bottom side enters the field and after that the magnetic force just balances mg.)

$$mg = \frac{(BL)^2}{R} v$$

(e) The velocity v is obtained by equating the change in potential energy, mgh, to the change in kinetic energy, $1/2 \, mv^2$. Thus $v^2 = 2gh$ or

$$v = (2gh)^{1/2}.$$

This leaves

$$mg = \frac{(BL)^2}{R} \sqrt{2\,gh}$$

an equation we can solve for h since all other parameters are given.

(f) Numerically we have

$$m = 3.6 \times 10^{-3} \text{ kg}, \quad L = 0.12 \text{ m}, \quad B = 0.5 \text{ T}, \quad R = 0.1 \, \Omega, \quad g = 9.8 \text{ m/s}^2$$

so that

$$h = 4.9 \times 10^{-2} \text{ m or } h = 4.9 \text{ cm}.$$

Example 9

Suppose the coil in the previous example is dropped with the shorter side, w, pointing down. If it is dropped from the same height as the other coil, how do the retarding forces compare?

Solution:

The retarding force on the first coil (F_1) was equal to

$$F_1 = \frac{(BL)^2 v}{R}$$

The retarding force on the second configuration of this coil (F_2) is found by replacing L by w so:

$$F_2 = \frac{(Bw)^2 v}{R}$$

Thus in this problem since $w = 1/2 \, L$ we have $F_2 = F_1/4$ and the coil, dropped with the short side down, would still experience a net downward force in the gravitational field. Its acceleration would be 3g/4 rather than zero and the braking effect of the magnetic field is greatly reduced. This is the general idea behind laminating metal parts that move in magnetic fields to reduce the eddy current effects. In this example, we reduced the length of the conductor carrying the current in the field and this in turn reduced the force considerably.

A practical consequence of Faraday's law is the existence of eddy currents in metallic pieces that move in magnetic fields. These currents can be beneficial and used as a "brake" on the motion or they can be troublesome and need to be reduced. In Example 9 a simple situation has been presented to illustrate these ideas.

Example 10

Use of Lenz's law, by trial and error, can determine the sign of the emf E_{nc}. In Figure 29-7 find the direction of E_{nc} for (a) an increase in magnetic field (ΔB points up) and (b) a decrease in magnetic field. (ΔB points down).

In the figure shown, a portion of a solenoid containing magnetic field lines is shown.

A larger loop encircles the region of magnetic field lines and an induced EMF, with counter clockwise direction is shown.

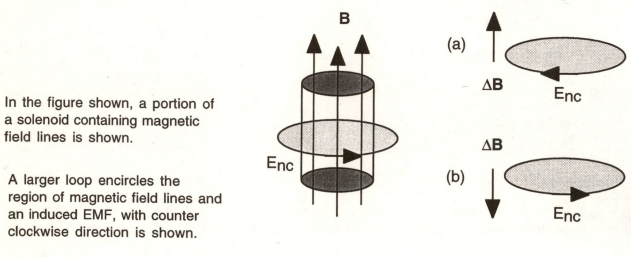

Figure 29-7

Solution

(a) Simply assume (trial and error) that the direction of Enc is counter-clockwise as shown in Fig. 29-7. We then test the consequences of that assumption--and if we find an error or inconsistency, we know we made the wrong choice.

(a) If the magnetic field increases (ΔB points up), the flux through the lightly shaded loop also increases and induces the EMF E_{nc}. An induced EMF produces a current in the direction of E_{nc} and that current produces an additional magnetic field. As drawn, using the right-hand rule, that current would produce a magnetic field <u>in the same direction as the initial increase in field</u>. This cannot be correct according to Lenz's law so the direction assumed for the EMF is wrong. The positive sense of E_{nc} for this case is clockwise as shown in the inset (a).

(b) If the original magnetic field decreases (ΔB points down), the flux through the lightly shaded loop decreases but still induces the EMF E_{nc}. We assume the direction of E_{nc} to be counter-clockwise (as shown) again. As drawn, using the right-hand rule, that resulting current would produce a magnetic field in the same direction as the initial field but in the opposite direction to the change in field. This is correct according to Lenz's law so the direction assumed for the EMF is right. The positive sense of E_{nc} for this case is counter-clockwise as shown in the inset (b).

QUIZ

1. A conductor of length L = 0.2 m, in a constant and uniform magnetic field, B = 3 T, slides over contacts at a constant velocity v = 0.3 m/s. The moving conductor is part of a complete circuit. The magnetic field is perpendicular to the plane of the circuit. Refer to Fig. 29-1. If the electrical resistance of the complete circuit is constant and equal to 0.4 Ω,
     (a) calculate the emf generated
     (b) calculate the current in the circuit
     (c) calculate the force on the moving conductor due to this motion in the magnetic field.

Answer: (a) emf = 0.18 volts; (b) I = 0.45 A; and (c) F = 0.27 N

2. A circular loop of wire (R = 0.15 m) rotates with constant angular velocity (ω = 100 π rad/s) in a constant and uniform magnetic field B = 1.8 T. The loop rotates about a diameter which is perpendicular to the magnetic field direction. Calculate the maximum emf generated in the coil.

Answer: emf = 40.0 volts

3. The velocity of a car traveling due east is constant and equal to 150 km/hr. If the earth's magnetic field is vertical at the location of the car and equal to 1.6×10^{-5} T (pointing downward), (a) calculate the magnitude of the emf induced in the axle of the car if the axle is 2 m long. (b) Use Lenz's law to find the direction of the emf.

Answer: (a) The emf = 1.33×10^{-3} volts. (b) A positive charge would be forced north.

4. A coil form of area 3 cm^2 has a total of 15 turns of wire wrapped on it. The leads of the coil are connected together through an external resistance of 1 MΩ. If the coil form, with its normal initially parallel to a constant magnetic field B = 3 T, is rotated through 180° so that its normal is now antiparallel to the field, calculate the magnitude of the charge that flows through the resistor.

Answer: Q = 2.7×10^{-8} C.

5. A stationary coil of radius r = 10 cm is placed so that its normal is parallel to a magnetic field B that is uniform in space (see Example 5 for geometry) but varies in time according to:

$$B = B_0 \cos [(377 \text{ rad/s})t]$$

where t is in seconds. Calculate the peak emf generated in this coil if $B_0 = 3 \times 10^{-3}$ T.

Answer: $V_{peak} = 3.55 \times 10^{-2}$ V.

30
INDUCTANCE

OBJECTIVES

In this chapter, inductance is introduced. Like capacitance, it is a property that depends on the geometric dimensions of the circuit element. Your objectives are to:

Calculate, for a particular geometry, both the *self-inductance* (L) and *mutual inductance* (M) of a circuit element or elements, from the definitions for L and M.

Calculate the energy stored in the magnetic field.

Analyze voltage-current relations in simple practical circuits containing self-inductances.

REVIEW

Inductance, like capacitance, is a physical property of components commonly used in electrical circuits. We associate inductance most frequently with wire, coils, wrappings, etc., but virtually all components used in circuits have some inductance. As capacitance could be used to store energy in the electric field so inductance can be used to store energy in the magnetic field established in the particular geometry.

Mutual inductance is a property of components which arises because the magnetic field produced by one element links other elements and can produce emf's in them by virtue of Faraday's law. This emf is thus directly related to the current change in the first component since magnetic fields result from current distributions. If $I_1(t)$ is the current in element (coil) 1, then the magnetic field at element 2 due to 1 can be calculated using the Biot law. Integration of this field over the area of element 2 gives the flux through one turn of element 2 due to the current distribution of element 1. If element 2 has N_2 turns, then according to Faraday's law, the emf induced in it by a current change in element 1 is:

$$\left(\text{emf}\right)_2 = -N_2 \frac{d\Phi_1}{dt} = -M_{12} \frac{dI_1}{dt}$$

where M_{12}, the mutual inductance of elements 1 and 2, simply relates this induced emf in element 2 to the current change in element 1. Since N_2 and M_{12} are independent of time, we have also:

$$N_2\Phi_1 = M_{12}I_1$$

These relationships are illustrated in Example 1. The mutual inductance of two elements depends only on their geometries and relative position and not upon which element is considered the source of the magnetic field. This statement is easy to prove for two interpenetrating solenoids (see Example 2) but much more difficult to prove for other current distributions.

An element or a coil, by itself, has inductance called self-inductance and denoted by the symbol L. This property is most easily discussed for a toroid but it is a general property of all components. In the case of a toroid, the current in the windings sets up a magnetic field which is constant over the volume of the toroid and zero elsewhere. For a toroid with N turns wound on a core of circumference L this field is approximately

$$B = \mu_0\left(\frac{N}{L}\right) I.$$

This field produces a flux through each turn of magnitude $\Phi_1 = BA$ where A is the cross-sectional area of the toroid. The emf induced in the toroid due to a current change would be:

$$(emf)_1 = - N_1 \frac{d\Phi_1}{dt}.$$

The self inductance L is defined by the equation

$$(emf)_1 = - L \frac{dI_1}{dt}.$$

Thus we can write (since N and L are independent of time)

$$N\Phi_1 = LI_1.$$

This is correct for all current distributions. For our specific example, the toroid, we have

$$L = \frac{N\Phi_1}{I} = \frac{N}{I} BA = \mu_0 \frac{N^2}{L} A.$$

Apart from the factor μ_0, both L and M are dimensionally lengths since the number of turns carries no dimensions. The self-inductance of virtually every configuration of circuit elements has been calculated and can be regarded as a tabulated quality which could be found if necessary from the Biot law.

Since the power input to a circuit element is the product of the emf and the current, for an inductance, either mutual or self, we have for the energy stored in an inductor:

$$\frac{dW}{dt} = LI \frac{dI}{dt} = \frac{d}{dt}\left(\frac{1}{2} LI^2\right)$$

Thus we can associate an energy of $1/2\, LI^2$ with an inductor. This energy is stored in the magnetic field surrounding the particular element of inductance L, carrying current I.

The unit for mutual and self-inductance is the henry. If a current change of one ampere per second produces an induced voltage of 1 volt, the inductance (self or mutual) is equal to one henry. Apart from the factor μ_0, the dimension of inductance is simply length. We should also note that since a volt per ampere is an ohm that a henry is also equal to an ohm-second. For this reason an inductance divided by a resistance has dimensions of time.

Since a current change in an inductance (or inductor) results in an emf, an inductance is an important practical circuit element. Its role in circuits containing a resistance, R, an R-L circuit and in circuits containing a capacitor and inductance (an L-C circuit) is discussed in the text. A series circuit containing all three elements (an R-L-C circuit) is discussed in a physical way but without detailed mathematical calculation.

For the R-L circuit, the transient behavior is obtained by recognizing that the sum of the emf's is equal to the sum of the potential drops (Kirchhoff's law). If V_0 represents a d.c. voltage which is suddenly impressed on a resistor connected in series with an inductor (for instance by closing a switch in the circuit), then the sum of the emf's is equal to $V_0 - LdI/dt$ whereas the sum of the potential drops is just equal to IR. Thus we have

$$V_0 - L\left(\frac{dI}{dt}\right) = IR,$$

a differential equation that is solved in the text to give I as a function of time. The solution is shown to be

$$I(t) = \frac{V_0}{R}\left(1 - e^{-t/\tau}\right)$$

where τ, the time constant, is equal to L/R and has the same dimensions as time. For very long times, the exponential becomes negligibly small and $I(t) = V_0/R$ as it would in a simple d.c. circuit containing a resistance R only. For short times $t \ll \tau$, the exponential can be approximated by

$$e^{-t/\tau} \cong 1 - t/\tau$$

so that

$$I(t) \cong \frac{V_0}{R}\frac{t}{\tau}.$$

Thus the current initially rises linearly with time (more slowly for larger L) and then for long times approaches the value V_0/R.

If we charge a capacitor (C) to an initial value Q_0 and then connect it to an inductor (L) by closing a switch, oscillation occurs just as with a mass vibrating on a perfect spring. We can use Kirchhoff's law once again to write:

$$- L \frac{dI}{dt} = V_{cap} = \frac{1}{C} q(t)$$

Since $I(t) = dq/dt$ we can rewrite this in the form of a second order differential equation

$$\frac{d^2q}{dt^2} + \frac{1}{LC} q(t) = 0 \ .$$

We wrote the spring equation characteristic of single harmonic motion with angular frequency ω_0 in the form

$$\frac{d^2x}{dt^2} + \omega_0^2 \, x = 0 \ ,$$

so that a comparison of the above two equations shows that the previous equation describes such a simple harmonic oscillation (of the charge on the capacitor and hence the current through the inductor) at an angular frequency

$$\omega_0 = \frac{1}{\sqrt{LC}}$$

Energy stored in the electric field of the capacitor (like the potential energy of the spring) is converted into energy stored in the magnetic field of the inductor (analogous to the kinetic energy of the mass m on the spring). No energy is lost or dissipated in this circuit.

If we add a resistive element, R, to this circuit, then for each cycle of oscillation of the circuit a certain amount of energy is lost by heat generated by the resistor. If the energy loss per cycle is small compared to the total energy in the circuit, the circuit continues to oscillate for many cycles at an altered frequency but the amplitude of the oscillation damps out and eventually the oscillation stops. If the energy loss per cycle is higher and roughly equal to the total energy in the circuit, the amplitude of the oscillation (here it is the stored charge initially on the capacitor) quickly goes to zero and no real oscillation may be discerned. A physical situation analogous to the series R-L-C circuit is found in many systems.

HINTS AND PROBLEM-SOLVING STRATEGIES

To calculate the mutual or self-inductance of a circuit or element, a very systematic procedure can be used:

(1) Use the Biot law or Ampere's law to find the magnetic field (at an arbitrary point) for the given current distribution. This gives B as a function of the position and the current I.

(2) Use the dependence of B on position to calculate the flux enclosed by the circuit or element.

(3) If there are multiple turns involved, equate the product of the number of turns and the flux through a single turn to the inductance multiplied by the current and then solve for the inductance.

In working circuit problems with inductances, include an inductance as a source of emf equal to -L(dI/dt) (or M if a mutual inductance is being considered) and then use Ohm's law. A pure (resistanceless) inductance does not dissipate any power.

QUESTIONS AND ANSWERS

Question. Given two flat circular coils, how would you arrange their relative orientation so that their mutual inductance was minimized? Maximized?

Answer. To minimize their mutual inductance, orient the coils so that their axes are perpendicular but for maximum mutual inductance, orient them so that their axes are parallel.

Question. Precision resistors are wound from a precisely determined length of wire. How would you wind a wire into the shape of a coil to get maximum inductance? Minimum inductance?

Answer. If the coil is tightly wound, single strand, on a cylindrical form so that the number of turns per meter is maximized, you will obtain the maximum inductance for a given wire diameter. To minimize the inductance, fold the wire in half before winding on the form so that the magnetic fields (and hence flux) from the two adjacent windings tend to cancel in the center of the coil. This is called an "astatic winding".

EXAMPLES AND SOLUTIONS

Example 1

Calculate the mutual inductance between a long straight wire and a rectangular circuit of dimensions 0.1 m by 0.2 m positioned with respect to the wire as shown in Fig. 30-1a, 30-1b, and 30-1c.

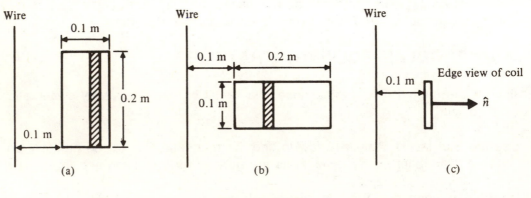

Figure 30-1

Solution:

(a) We assume there is a current I_1 in the long straight wire since this is the easiest magnetic field to calculate in this problem and calculate the flux in the loop due to this current . If we measure the coordinate r radially from the wire, we have

$$B_1 = \frac{\mu_0}{4\pi} \frac{2I_1}{r} .$$

The element of flux, $d\Phi_2$, through the cross-hatched element of area, $dA = L\,dr$ (where L is the length of coil parallel to the wire) is:

$$d\Phi_2 = \frac{\mu_0}{4\pi} 2I_1 L\left(\frac{dr}{r}\right) .$$

If we integrate dr/r from r_1 to r_2 we obtain the flux through element 2 (the rectangular coil) due to the current in element 1 (the wire);

$$\Phi_2 = \frac{\mu_0}{4\pi} 2I_1 L \int_{r_1}^{r_2}\left(\frac{dr}{r}\right) = \frac{\mu_0}{4\pi} 2I_1 L \ln\left(\frac{r_2}{r_1}\right)$$

Since the rectangular coil has only one turn, $\Phi_2 = M_{12}I_1$ and we have:

$$M_{12} = \frac{\mu_0}{4\pi} 2L \ln\left(\frac{r_2}{r_1}\right)$$

Numerically, for part (a), we have

$$M_{12}^{(a)} = (10^{-7})(0.4) \ln 2 \text{ henry.}$$

$$= 2.77 \times 10^{-8} \text{ henry or } 0.0277 \ \mu H$$

(b) The general formula developed in (a) can still be used here but the numerical value is changed:

$$M_{12}^{(b)} = (10^{-7})(0.2) \ln 3 \text{ henry}$$

$$= 2.20 \times 10^{-8} \text{ henry.}$$

(c) For this part, since $\vec{B}_1 \cdot \hat{n} = 0$ due to the orientation of the coil, the flux Φ_2 is zero and $M_{12}^{(c)} = 0$. From parts a, b, and c we can conclude that the mutual inductance depends critically on the relative orientation of the elements.

Example 2

Consider a short coil (length L_1) of radius R_1 with N_1 turns inside a second *long* (length $L_2 >> L_1$) solenoid of radius R_2 with N_2 turns. Calculate the mutual inductance of the pair
     (a) if the axes of the two coils are parallel
     (b) if the axes of the two coils are perpendicular.

Solution:

(a) Assume a current I_2 flows through the outer solenoid. This gives a uniform field

$$B_2 = \mu_0 \frac{N_2}{L_2} I_2$$

throughout the region interior to this solenoid and near its center. The flux through the inner solenoid due to I_2 is then B_2 multiplied by the cross-sectional area of the inner solenoid (πR_1^2).

$$\Phi_1 = \mu_0 (N_2/L_2) I_2 (\pi R_1^2)$$

Since the inner solenoid has N_1 turns, we write

$$N_1 \Phi_1 = M_{21} I_2$$

and solving for M_{21} we find that

$$M_{21} = \frac{N_1 \Phi_1}{I_2} = \mu_0 \frac{N_1 N_2}{L_2} \pi R_1^2$$

(b) If the axes are perpendicular, the flux through coil 1 is zero as the normal to its surface is perpendicular to $\vec{B}_2$. The mutual inductance is now zero. Although the coils are the same, their geometric arrangement is different so the mutual inductance is different.

Example 3

The capacitor shown in Fig. 30-2 initially holds a charge Q_0 and no current flows until the switch S is closed. Obtain the charge and the current as functions of time.

(a) Evaluate q and i for t = $(\pi/2) \times 10^{-4}$ s and t = $\pi \times 10^{-4}$ s if $Q_0 = 10^{-9}$ C, C = 2 μF, and L = 5 mH.

(b) Verify that the energy is conserved.

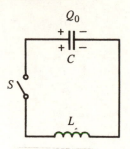

Figure 30-2

Solution:

(a) After S is closed, the inductance L acts as a source of emf equal to -Ldi/dt. The voltage drop across the capacitor, V_C, is equal to q/C so Ohm's law gives:

$$-L\left(\frac{di}{dt}\right) = \left(\frac{1}{C}\right)q$$

Since i = dq/dt, di/dt = d^2q/dt^2 we can rearrange the above equation as

$$\frac{d^2q}{dt^2} + \frac{1}{LC}\, q(t) = 0 \ .$$

This is to be compared with the equation we obtained for the displacement of a mass m on a spring (with force constant k):

$$\frac{d^2x}{dt^2} + \frac{k}{m}\, x = \frac{d^2x}{dt^2} + \omega_0^2\, x = 0$$

Thus $\omega_0^2 = 1/LC$ for the present case. The general solution for the charge is

$$q = A \cos \omega_0 t + B \sin \omega_0 t$$

and the current is

$$i = \frac{dq}{dt} = -\omega_0 A \sin \omega_0 t + \omega_0 B \cos \omega_0 t$$

To find A and B we note that at $t = 0$, we have $i = \omega_0 B$. Since $i = 0$ at $t = 0$ then B must be zero. At $t = 0$, we have $q = A$ but since the initial charge is specified as Q_0, then $A = Q_0$ leaving

$$q = Q_0 \cos \omega_0 t,$$

and

$$i = - \omega_0 Q_0 \sin \omega_0 t.$$

For this problem

$$\omega_0^2 = \frac{1}{LC} = \frac{1}{(5 \times 10^{-3} \text{ H})(2 \times 10^{-6} \text{ F})} = 10^8 \text{ s}^{-2}$$

Thus $\omega_0 = 10^4$ s^{-1}.

At $t = 0$, $q = Q_0$ and $i = 0$. At $t = (\pi/2) \times 10^{-4}$ s, we have

$$\omega t = (10^4 \text{ s}^{-1})(\pi/2)(10^{-4} \text{ s}) = \pi/2$$

and

$$q = Q_0 \cos (\pi/2) = 0$$

$$i = - \omega Q_0 \sin (\pi/2) = - \omega Q_0 = - 10^{-5} \text{ A}$$

At $t = \pi \times 10^{-4}$ s, we have

$$q = Q_0 \cos (\pi) = - Q_0 = - 10^{-9} \text{ C}$$

$$i = - \omega Q_0 \sin (\pi) = 0.$$

(b) The energy stored in the capacitor is given by

$$E_C = \frac{1}{2}\left(\frac{q^2}{C}\right) = \frac{1}{2C} Q_0^2 \cos^2 \omega_0 t$$

The energy stored in the inductor is:

$$E_L = \frac{1}{2}\left(Li^2\right) = \frac{1}{2} L\omega_0^2 Q_0^2 \sin^2 \omega_0 t$$

$E_C + E_L$ is constant since $1/C = L\omega_0^2$ and $\sin^2 \omega_0 t + \cos^2 \omega_0 t = 1$.

Example 4

Obtain the equivalent single inductance, L_E, to replace two inductances L_1 and L_2 connected (a) in series, and (b) in parallel. See Fig. 30-3a and b.

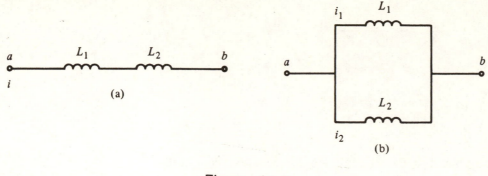

Figure 30-3

Solution:

(a) For two inductances in series, the current i through them is the same and the emf developed between points a and b is equal to:

$$V_{ab} = - L_1 \frac{di}{dt} - L_2 \frac{di}{dt} = - (L_1 + L_2) \frac{di}{dt} .$$

The same emf would be obtained if L_1 and L_2 were replaced by one inductance, L_E provided $L_E = L_1 + L_2$.

(b) For the parallel combination, we want to find L_E such that

$$V_{ab} = - L_E \, di/dt.$$

We have that $V_{ab} = - L_1 \, di_1/dt$ and also that $V_{ab} = - L_2 \, di_2/dt$. Since $i_1 + i_2 = i$, replace i_2 by $i - i_1$, and write:

$$V_{ab} = - L_1 \frac{di_1}{dt} = - L_2 \frac{d}{dt} (i - i_1) ;$$

Collecting both terms involving i_1 gives

$$- (L_1 + L_2) \frac{di_1}{dt} = - L_2 \frac{di}{dt} .$$

or by rearranging

161

$$- \frac{di_1}{dt} = - \left(\frac{L_2}{L_1 + L_2} \right) \frac{di}{dt} .$$

Multiply by L_1 to obtain

$$V_{ab} = - L_1 \frac{di_1}{dt} = - \left(\frac{L_1 L_2}{L_1 + L_2} \right) \frac{di}{dt} = - (L_E) \frac{di}{dt} .$$

Thus

$$L_E = \left(\frac{L_1 L_2}{L_1 + L_2} \right) .$$

This is exactly like the result obtained for resistors in parallel:

$$\frac{1}{L_E} = \frac{1}{L_1} + \frac{1}{L_2} .$$

Example 5

Calculate the self-inductance per unit length of a pair of concentric thin cylinders of radii R_1 and R_2 where $R_2 > R_1$ by calculating the magnetic flux in the region between the two cylinders. See Fig. 30-4a, b.

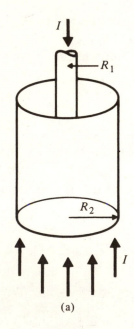

(a)

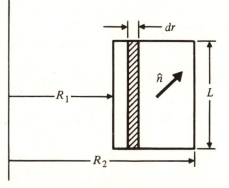

(b)

Figure 30-4

Solution:

Before working this problem, review Example 6 in Chapter 29 where the magnetic field B was calculated for this geometry. There it was found that for equal but opposite currents, I, on the two surfaces, that

(a) $B_1 = \dfrac{\mu_0}{4\pi} \dfrac{2I_1}{r}$ for $R_1 < r < R_2$.

This value comes just from the current on the inner cylinder because B due to current on the outer cylinder is zero.

(b) B = 0 outside the pair of cylinders (since the *net* current through the Amperian surface is zero).

Thus choosing a surface as shown in Fig. 30-4b with $\hat{n}$ pointing out of the paper, the flux $d\Phi$ through the cross-hatched region (B is constant over this area) is

$$d\Phi = \vec{B} \cdot \hat{n}\, dA$$

$$d\Phi = \dfrac{\mu_0}{4\pi} \dfrac{2I}{r} Y\, dr,$$

so by integrating

$$\Phi = \dfrac{\mu_0}{4\pi}\, 2I\, Y \int_{R_1}^{R_2} \dfrac{dr}{r},$$

$$\Phi = \dfrac{\mu_0}{4\pi}\, 2I\, Y \ln\left(\dfrac{R_2}{R_1}\right).$$

Since we have only "one turn" involved here, $\Phi = LI$ giving for the inductance per unit length, L/Y the value;

$$\dfrac{L}{Y} = \dfrac{\Phi}{IY} = \dfrac{\mu_0}{4\pi}\, 2 \ln\left(\dfrac{R_2}{R_1}\right).$$

Example 6

Calculate the self-inductance per unit length of a pair of concentric thin cylinders of radii R_1 and R_2 where $R_2 > R_1$ by equating the energy stored per unit volume in the magnetic field in the region between the two cylinders to the energy stored in an inductance. See Fig. 30-4a, b.

163

Solution

The energy (U) stored in an inductance is given by:

$$U = \frac{1}{2} LI^2$$

The energy stored by the magnetic field, per unit volume, u, is given by:

$$u = \frac{1}{2\mu_0} B^2 = \frac{U}{V}$$

Since B is not constant over the region between the two cylinders, we must integrate to find U.

$$dU = \frac{1}{2\mu_0} B^2 dV .$$

Let $dV = 2\pi r \, l \, dr$ and

$$B_1 = \frac{\mu_0}{4\pi} \frac{2I_1}{r} .$$

as before. Then we have,

$$dU = \frac{1}{2\mu_0} \left(\frac{\mu_0}{4\pi} \frac{2I}{r} \right)^2 2\pi r Y \, dr$$

$$dU = \frac{\mu_0}{4\pi} Y I^2 \frac{dr}{r} .$$

Integrating from R_1 to R_2, we obtain U for the magnetic field--which we then equate to the quantity $LI^2/2$:

$$U = \frac{\mu_0}{4\pi} Y I^2 \ln \left(\frac{R_2}{R_1} \right) = \frac{1}{2} LI^2 .$$

so

$$L = \frac{\mu_0}{4\pi} 2Y \ln \left(\frac{R_2}{R_1} \right) \quad \text{as before.}$$

Example 7

Suppose we connect a "lossy" or real inductor (inductance L and resistance r) in series with a light bulb which we represent by a pure resistance R as shown in Fig. 30-5.

(a) Calculate the voltage drop, V, across the light bulb after the switch S is closed.

(b) Assume the switch S has been closed a long time so that there is a steady current. If switch S is now opened, calculate V as a function of time.

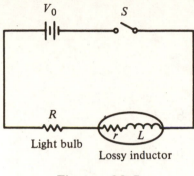

Figure 30-5

Solution:

(a) After S is closed, we regard the inductance as a source of emf and use Ohm's law:

$$V_0 - L(di/dt) = (R + r)i$$

The variables, i and t are separable here so this equation can be integrated as done in the text. For variety, we can solve this by a change of variables. Write

$$\frac{di}{dt} = -\left(\frac{R + r}{L}\right)\left(i - \frac{V_0}{R + r}\right).$$

Letting

$$x = \left(i - \frac{V_0}{R + r}\right)$$

we have

$$\frac{dx}{dt} = \frac{di}{dt}. \qquad \text{Set } \tau = \frac{L}{R + r}.$$

Then we have

$$\frac{dx}{dt} = -\frac{1}{\tau}x \qquad \text{or } x = x_0\, e^{-t/\tau}.$$

At t = 0, there is no current, so

$$x_0 = -\left(\frac{V_0}{R + r}\right)$$

giving

$$i - \left(\frac{V_0}{R + r}\right) = -\left(\frac{V_0}{R + r}\right) e^{-t/\tau}$$

or

$$i = \left(\frac{V_0}{R + r}\right)\left(1 - e^{-t/\tau}\right).$$

The voltage drop across the light bulb V is:

$$V = i(t)R = \left(\frac{V_0 R}{R + r}\right)\left(1 - e^{-t/\tau}\right).$$

Thus the bulb is turned on smoothly and the larger the ratio $L/(R + r) = \tau$, the slower the voltage increase.

(b) For very long times, the current reaches the value $i = V_0/(R + r)$. If the switch S is now opened, Kirchhoff's law reads (since V_0 is no longer in the circuit as an emf),

$$-L\left(\frac{di}{dt}\right) = (R + r)\, i$$

which we readily integrate as

$$\frac{di}{i} = -\frac{dt}{\tau}$$

to give:

$$i(t) = i(0)e^{-t/\tau}.$$

For i(0) we use the value

$$i(0) = \frac{V_0}{R + r}$$

so the voltage across the light, V, decays exponentially to zero:

$$V = \left(\frac{V_0 R}{R + r}\right)\left(e^{-t/\tau}\right).$$

QUIZ

1. A toroid carrying a current of 3 A has a self inductance of 1 mH. The mean radius of the toroid is 0.075 m and the cross-sectional area is equal to 5×10^{-4} m^2.
     (a) Calculate the number of turns on the toroid.
     (b) Calculate the flux through the toroid.
     (c) Calculate the energy stored in the toroid.

Answer: (a) $N = 7.5 \times 10^5$;   (b) flux $= 2.25 \times 10^6$ webers; (c) $E = 4.5 \times 10^{-3}$ J

2. A very long solenoid with radius R = 0.08 m has 4000 turns per meter. A second coil consisting of 150 turns is wound around the center of the solenoid. Calculate the mutual inductance of this pair.

Answer: $M = 1.52 \times 10^{-2}$ H.

3. A toroid of cross-section A = 5×10^{-4} m^2 and mean radius R = 0.2 m carries a current of 5 A. The toroid has 7000 turns. (a) Assuming the magnetic field inside the toroid is uniform and equal to the value along the circumference, calculate the energy per unit volume stored in the toroid.
(b) Using the expression for the energy stored in an inductance and the result from part (a), calculate the self inductance of the toroid.

Answer: (a) $u = 487.4$ J/m^3; (b) $L = 2.45 \times 10^{-2}$ H.

4. A 5 H inductor carrying an initial current of 100 A is discharged through a resistor of 1 Ω.
(a) Calculate the initial voltage drop across the resistor and (b) the time needed for the current to reach a value of 10 A.

Answer: (a) $V = 100$ volts; (b) t = 11.5 s.

5. Coil #2 is wrapped closely around coil #1. When the current in coil #1 changes at the rate of 3 A in 0.001 s, the voltage induced in coil #2 is 9 mV. If a current in coil #2 changes at the rate of 2000 A/s, what voltage is induced in coil #1?

Answer: 6 mV.

31

ALTERNATING CURRENT

OBJECTIVES

In this chapter you will define inductive and capacitive reactance, quantities that like resistance limit the current flow for given difference in potential. A method is given for analyzing simple circuits subject to sinusoidal voltages at a single frequency and consisting of passive elements R, L and C. Your objectives are to:

Obtain the source voltage as the *vector sum* of voltages across the individual components in a series R-L-C circuit.

Obtain the source current as a *vector sum* of currents through the individual elements in a parallel R-L-C circuit.

Calculate the phase difference between source voltage and current.

Calculate the time average power consumed by the circuit.

REVIEW

We have previously studied electrical circuits where both steady currents and time dependent currents were present. The passive circuit elements, resistance, capacitance, and inductance have each been studied separately so that we know how to relate the potential difference across each element to the current, the integral of the current with respect to time (the charge) or the time derivative of the current. Thus whenever the current is not constant, we generally must solve a differential equation. We now will study the special (but very important) case where the voltage source produces an emf that is a sinusoidal (sine or cosine) function of time at a single frequency, f. We have

$v = V \cos \omega t$

where (lower case) v is the <u>instantaneous</u> value of the emf and (upper case) V is constant and equal to the <u>maximum</u> value of the emf. The angular frequency is $\omega = 2\pi f$. The general solution to the problem of connecting such a source to a circuit will consist of a transient current, as we found previously for inductors and capacitors, plus a sinusoidal current at the same frequency as the source voltage. We will ignore the transient behavior in this chapter and concentrate on the sinusoidal solution.

If a source of voltage $v = V \cos \omega t$ is connected to a resistance element, R, the current is

$$i(t) = \frac{v(t)}{R} = \frac{V \cos \omega t}{R} = I \cos \omega t$$

where the maximum value of current is $I = V/R$. Because the current and voltage have precisely the same time dependence, they are said to be "in phase".

If, instead, a capacitor, C, is connected to this alternating voltage source, the charge on the capacitor oscillates sinusoidally since $q(t) = Cv(t)$. The instantaneous current through the capacitor, $i(t)$, is equal to dq/dt so:

$$i(t) = \frac{dq}{dt} = \frac{d}{dt}\left(CV \cos \omega t\right) = -\omega CV \sin \omega t = -I \sin \omega t$$

The maximum value of this current is, $I = \omega CV$. In analogy to DC circuit analysis, the maximum current can be thought of as V/X_C where X_C, the <u>capacitive reactance</u>, is equal to $1/\omega C$ and plays the role of a resistance because it relates a maximum current to the maximum voltage. The unit for X_C is the ohm. The above current does not have the same time dependence as the voltage; they are "out of phase". We can be quantitative about this by using the identities:

$$\cos(\omega t + \phi) = \cos \omega t \cos \phi - \sin \omega t \sin \phi$$

$$\sin(\omega t + \phi) = \sin \omega t \cos \phi + \cos \omega t \sin \phi$$

Thus $-\sin \omega t$ is the same as $\cos(\omega t + 90°)$. We conclude that for a capacitor the voltage and current are 90° "out of phase", with the current *leading* the voltage.

For an inductor, L, connected (alone) to the same alternating voltage source, since

$$v(t) = L \frac{di}{dt} ;$$

by integrating di/dt, we have

$$i(t) = \left(\frac{V}{\omega L}\right) \sin \omega t.$$

Technically since we integrated $v(t)$ to get i, we should include a constant term in $i(t)$ but since the time average of i must be zero, the constant is zero. Here the maximum value of i is given by V/X_L where X_L, the <u>inductive reactance</u> is equal to ωL. Again the unit for X_L is the ohm. Note that $\sin \omega t = \cos(\omega t - 90°)$ so the current *lags* the voltage in an inductor. This is physically reasonable because of Lenz's law.

If we connect two or more of these elements in series to the a.c. voltage source, the source voltage, $v(t)$, will be equal to the algebraic sum of the voltage drops across the various components. For practical problems however it is preferable to write that the common

169

current, through the elements is

$$i(t) = I \sin \omega t$$

and calculate the voltage by *vector addition*. This can be illustrated by treating an RLC series combination. The algebraic solution is given in Example 5.

For the resistor,

$$v_R = Ri = RI \sin \omega t,$$

in phase with $i(t)$. For the inductor

$$v_L = \left(\frac{di}{dt}\right) = \omega LI \cos \omega t = IX_L \sin(\omega t + 90°).$$

For the capacitor

$$v_C = \left(\frac{1}{C}\right) q(t) = \left(\frac{1}{C}\right) \int i \, dt = -\left(\frac{I}{\omega C}\right) \cos \omega t$$

$$= IX_C \sin(\omega t - 90°).$$

Referring to Fig. 31-1, let ωt be an angle measured c.c.w. (counter-clockwise) from the x axis. Thus if the x axis is used as the reference axis for the current, the voltage across the resistor is a vector (of magnitude IR) along the + x axis. The voltage across the inductor is a vector (of magnitude IX_L) along the + y axis. The voltage across the capacitor is a vector (of magnitude IX_C) along the - y axis.

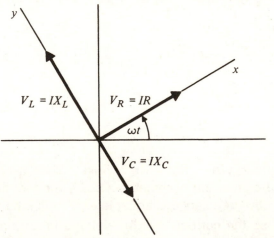

Figure 31-1

170

Using the rules for vector addition, the resultant voltage magnitude which is the magnitude of the voltage is found to be:

$$V_s = \sqrt{\left[(IR)^2 + (IX_L - IX_C)^2\right]} = IZ$$

The quantity Z introduced here is called the <u>impedance</u>. It is the single quantity characteristic of the whole circuit that is the ratio of the maximum voltage to the maximum current. Numerical solutions are given in Examples 2 and 3.

Regarding the source voltage as a vector quantity, it is seen that V lies above the + x axis by an angle ϕ where $\tan \phi = (X_L - X_C)/R$. If $X_C > X_L$, V lies below the + x axis. The angle ϕ thus gives the phase difference between the voltage and current in this series circuit such that if

$i = I \sin \omega t$

then

$v = IZ \sin (\omega t + \phi)$ where $Z = (R^2 + (X_L - X_C)^2)^{1/2}$.

The unit for Z is the ohm.

The instantaneous power is the product of i(t) with v(t). For the circuit considered, this is equal to

$P(t) = I^2Z \sin \omega t \sin(\omega t + \phi).$

Using the general expression for time averages in the text (see Example 1) and expanding $\sin(\omega t + \phi)$ we find that the time average power, P_{avg}, is:

$$P_{avg} = \frac{1}{2} I^2Z \cos \phi.$$

Since $Z \cos \phi = R$, then $P_{avg} = (1/2)I^2R$, which is the power consumed by the resistor alone. This is not surprising since neither the capacitor nor the inductor consume any power (on time average). The similarity of this expression with that obtained with d.c. circuits leads to the definition of r.m.s. quantities (a.c. instruments are calibrated to read these "root-mean-square" quantities *not* the maximum values).

$$P_{avg} = \left(I_{rms}\right)^2R = \frac{1}{2} I^2R$$

from which we conclude that

$$I_{rms} = \frac{1}{\sqrt{2}} I.$$

The phenomenon called "resonance" in such a series R-L-C circuit occurs when $X_L = X_C$ so the impedance Z is equal to R. For a given source voltage, the maximum current is obtained under these conditions as Z is minimum here. Note also that the phase angle ϕ is zero and the source voltage and current are in phase. Maximum power is drawn by the circuit on resonance and the "power factor", $\cos \phi$, is unity. See Example 4. Resonance in parallel circuits is treated briefly in Example 6.

An extremely practical device employed in a.c. circuits and invaluable for power transmission is the transformer. If an iron (or equivalent) core is used, the flux (Φ) linking the primary and secondary portions is the same. If the secondary circuit is open, we have

$$\mathcal{E}_p = - N_p \frac{d\Phi}{dt} \quad \text{and} \quad \mathcal{E}_s = - N_s \frac{d\Phi}{dt}$$

so

$$\frac{\mathcal{E}_p}{N_p} = \frac{\mathcal{E}_s}{N_s} .$$

This enables the transformer to either provide a secondary voltage greater than the primary voltage (step-up transformer $N_s > N_p$) or to provide a secondary voltage less than the primary voltage (step-down transformer $N_s < N_p$). When the secondary circuit is closed and draws current, the primary current must change to keep the flux change fixed. Since the *primary* impedance is the ratio of the primary voltage to the primary current, this impedance is determined by the secondary impedance and the turns ratio of the transformer. For this reason, changes in the secondary are reflected in the primary. See Example 7.

PROBLEM-SOLVING STRATEGY

Frequently, in this chapter, the problems are more of a numerical type. For such problems, if the angular frequency is known and the values of L and C are known, calculate the reactances numerically.

Next provide a sketch with the resistance plotted along the + x axis, the inductive reactance plotted along the + y axis, and the capacitive reactance plotted along the – y axis. Add, algebraically, the inductive reactance and capacitive reactance; this is the y component of the impedance. Now add, as vectors, the resistance R as the x component and the net y component to find the impedance Z. The phase angle ϕ can be found from R and Z since $\cos \phi = R/Z$.

QUESTIONS AND ANSWERS

Question. For an RLC circuit, with fixed R and L, but variable C, how can you get minimum impedance?

Answer. For minimum impedance, arrange the components in a series circuit and vary C so that $X_C = X_L$ (the circuit is in resonance).

Question. It has been said that a series L-C circuit is analogous to a mass on a spring executing simple harmonic motion. How can this analogy be correct?

Answer. If the circuit is analyzed from the standpoint of energy exchanges, the energy stored in the coil $U_L = (1/2)LI^2$ is energy stored in the magnetic field while the energy stored in the capacitor $U_C = (1/2)Q^2/C$ is energy stored in the electric field. Since $dQ/dt = I$, the charge Q is analogous to the displacement of the mass on the spring while the current is analogous to the particle velocity. Thus the energy in the coil is like the kinetic energy while the energy in the capacitor is like potential energy. Energy is continuously converted from potential to kinetic (and vice versa) for the mass on the spring and the electric and magnetic fields.

EXAMPLES AND SOLUTIONS

Example 1

Calculate the time average of $\sin^2 \omega t$ over one period of the motion without doing any integrations.

Solution:

We have

$$1 = \sin^2 \theta + \cos^2 \theta = \sin^2 \omega t + \cos^2 \omega t$$

Over one cycle of $\sin \omega t$, the area under the curve $\sin^2 \omega t$ is equal to the area under the curve of $\cos^2 \omega t$. *Sketch this to be sure.* Thus if we use <f> to denote the time average of a quantity f over one complete cycle,

$$<\sin^2 \omega t> = <\cos^2 \omega t>$$

$$<1> = 1 = <\sin^2 \omega t> + <\cos^2 \omega t> = 2<\sin^2 \omega t>$$

We conclude that:

$$<\sin^2 \omega t> = (1/2) = <\cos^2 \omega t>$$

Example 2

A resistor and inductor as shown in Fig. 31-2a are connected in series to a 60 cycle voltage source which produces a maximum voltage V = 155.6 volts. If R = 200 Ω and L = 0.3 H, calculate
  (a) the impedance of the circuit;
  (b) the maximum current, I;
  (c) the phase angle between the current and the voltage; and
  (d) the time average power drawn from the source.

173

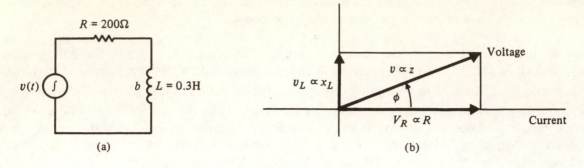

(a)   (b)

Figure 31-2

Solution:

(a) If the source frequency is 60 Hz. then $\omega = 2\pi(60\ s^{-1}) = 377$ rad/s. For this angular frequency, the inductive reactance, X_L, is:

$$X_L = \omega L = (377)(0.3)\ \Omega = 113.1\ \Omega$$

The impedance $Z = (R^2 + X_L^2)^{1/2}$. For $R = 200\ \Omega$ we have $Z = 230\ \Omega$ by direct substitution of these values. Note that Z is the magnitude of the vector that has R as its x component and X_L as its y component.

(b) The maximum current is related to the maximum voltage by:

$$V = IZ$$

Thus

$$I = \frac{V}{Z} = \frac{(155.6\ V)}{230\ \Omega} = 0.676\ \text{amps}$$

(c) From Fig. 31-2b, the resultant voltage (the source voltage) leads the current in the circuit by the angle ϕ where $\cos \phi = R/Z$. For this case $\phi = 29.6°$.

(d) The voltage across the resistor,v_R, is just Ri. So the instantaneous power through the resistor is $Ri^2 = RI^2 \sin^2 \omega t$. The time average value of $\sin^2\omega t$ is 1/2 so the time average power consumed by the resistor is:

$$P_R = (1/2)\ RI^2$$

For the inductor,

$$v_L = L(di/dt) = \omega LI \cos \omega t.$$

The instantaneous power, $P_L(t)$, is just

$$P_L(t) = v_L i = \omega L I^2 \sin \omega t \cos \omega t.$$

Since

$$\sin \omega t \cos \omega t = (1/2) \sin 2\omega t$$

and all sines and cosines give zero time average over a cycle, the time average power drawn by the inductor is *zero*.

$$<P_L(t)> = 0.$$

Finally, for the entire circuit, since

$$i = I \sin \omega t \text{ and } v = V \sin(\omega t + \phi),$$

the instantaneous power is:

$$iv = IV \sin \omega t \sin(\omega t + \phi).$$

If we write

$$\sin(\omega t + \phi) = \sin \omega t \cos \phi + \cos \omega t \sin \phi,$$

then since the time average of $\sin^2 \omega t$ is 1/2 and the time average of $\sin \omega t \cos \omega t$ is zero, we have:

$$P = (1/2) IV \cos \phi.$$

However $\cos \phi = R/Z$ and $V = IZ$ so this gives $P = (1/2) RI^2$, just the result obtained for the resistor alone.

Example 3

For the series R-L-C circuit shown in Fig. 31-3 calculate the maximum current and the phase difference between the voltage and the current if the maximum source voltage is 50 V and the frequency is
  (a) 5000 Hz and
  (b) 2000 Hz.

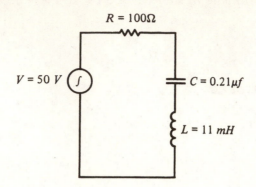

$R = 100\Omega$

$V = 50\ V$

$C = 0.21\mu f$

$L = 11\ mH$

Figure 31-3

Solution:

(a) If f = 5000 Hz, $\omega = 3.14 \times 10^4$ rad/s. The necessary reactances are

$$X_C = 1/\omega C = 1/[(3.14 \times 10^4\ s^{-1})(2.1 \times 10^{-7}\ F)] = 151.6\ \Omega$$

$$X_L = \omega L = (3.14 \times 10^4\ s^{-1})(11 \times 10^{-3}\ H) = 345.6\ \Omega$$

so the impedance Z has the value:

$$Z = [(100)^2 + (345.6 - 151.6)^2]^{1/2}\ \Omega = 218.3\ \Omega.$$

Since $X_L > X_C$, the voltage will lead the current by angle ϕ where

$$\cos \phi = (R/Z) = 0.458; \qquad \phi = 62.7°.$$

The maximum current is

$$I = \frac{V}{Z} = \frac{(50\ V)}{218\ \Omega} = 0.229\ A.$$

Thus if,

$$i(t) = (0.229\ A)\ \sin \omega t$$

then

176

$$v(t) = (50 \ V) \ \sin(\omega t + 62.7°)$$

(b) Let $f = 2000$ Hz so the values of the inductive and capacitive reactances are changed.

$$\omega = 1.257 \times 10^4 \ \text{rad/s}$$

$$X_L = \omega L = 138.2 \ \Omega$$

$$X_C = (\omega C)^{-1} = 378.9 \ \Omega$$

Now the capacitive reactance is larger than the inductive reactance so the voltage will lag behind the current by an angle ϕ.

$$Z = (R^2 + (X_L - X_C)^2)^{1/2}$$

$$= [(100)^2 + (138.2 - 378.9)^2]^{1/2} \ \Omega$$

$$= 261 \ \Omega$$

Since $\tan \phi = (X_L - X_C)/R$, then $\phi = -67.44°$. The maximum current is

$$I = \frac{V}{Z} = \frac{(50 \ V)}{(261 \ \Omega)} = 0.192 \ A.$$

Thus we have

$$i(t) = (0.192 \ A) \ \sin \omega t$$

$$v(t) = (50 \ V) \ \sin (\omega t - 67.4°).$$

Example 4

For a series R-L-C circuit carrying current $i = I \sin \omega t$, driven by a voltage source $v = V \sin(\omega t + \phi)$ where V is constant, find the angular frequency ω for which the time average power is a maximum.

Solution:

If we start with the expression for the time average power $P_{avg} = <P>$,

$$<P> = (1/2) \ V \ I \ \cos \phi,$$

we can replace $\cos \phi$, the power factor, by $\cos \phi = R/Z$ and I by V/Z to get:

$$\langle P \rangle = \frac{1}{2} \left(\frac{V}{Z} \right)^2 R.$$

For the impedance Z, use $Z^2 = R^2 + (\omega L - 1/\omega C)^2$. Since V and R are constants, the maximum power will be obtained when Z is a minimum. This occurs at a frequency ω_0 where

$$\omega_0 L = \frac{1}{\omega_0 C} \quad \text{or} \quad \omega_0^2 = \left(\frac{1}{LC} \right)$$

At this frequency $Z = R$ and the maximum power, P_{max}, is:

$$P_{max} = \frac{1}{2} \frac{V^2}{R}.$$

Example 5

For a series R-L-C circuit carrying current $i = I \sin \omega t$, obtain the voltage across the circuit using the sum of the algebraic expressions for the individual voltages.

Solution:

Writing the source voltage, v(t), as the algebraic sum of the voltage drops,

$$v(t) = v_R(t) + v_L(t) + v_C(t)$$

where the values of v_R, v_L, and v_C are given by:

$$v_R(t) = R\ i(t) = R\ I \sin \omega t$$

$$v_L(t) = L(di/dt) = \omega L\ I \cos \omega t$$

$$v_C(t) = (1/C)\ q(t) = - (I/\omega C) \cos \omega t$$

After substitution, we have

$$v(t) = I[(\omega L - 1/\omega C) \cos \omega t + R \sin \omega t]$$

Now define

$$\tan \phi = \left(\frac{[\omega L - 1/\omega C]}{R} \right)$$

so

$$\cos \phi = \frac{R}{\sqrt{[\omega L - 1/\omega C]^2 + R^2}} = \frac{R}{Z}$$

Then

$$v(t) = IR\left(\frac{\sin \phi}{\cos \phi} \cos \omega t + \sin \omega t\right)$$

$$v(t) = \frac{IR}{\cos \phi}(\sin \phi \cos \omega t + \cos \phi \sin \omega t)$$

$$= IZ \sin(\omega t + \phi).$$

The totally algebraic solution gives the same result as that obtained by vector addition but requires inspired guesses at a few spots (like the definition of tan ϕ).

Example 6

Suppose the components in Example 3 are placed in a parallel connection, with the same voltage source operating at 5000 Hz. What is the maximum value of the current?

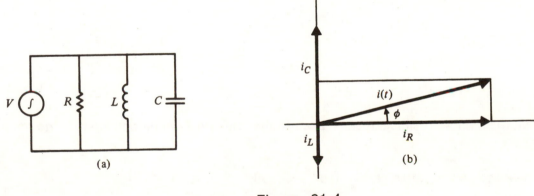

Figure 31-4

Solution:

Referring to Fig. 31-4a, use R = 100 Ω, L = 11 mH, and C = 0.21 μF with ω = 3.141 x 10^4 rad/s. In a parallel circuit, the potential difference across all the elements shown is the same but the current drawn from the source is the sum of the currents through R, L, and C, according to Kirchhoff's law. The *currents* (but not the voltages) have phase differences and must be added as vectors. Since v(t) is common to all the elements, we can plot i_R, the current through the resistor, along the + x axis in Fig. 31-4b. Since the current leads the voltage for a capacitor, we plot i_C along the + y axis and i_L (since the voltage leads the current in an inductor) along the – y axis. The magnitudes of these currents are:

$$i_R = \frac{V}{R} = \frac{(50 \text{ V})}{(100 \text{ }\Omega)} = 0.5 \text{ A}$$

$$i_C = \frac{V}{X_C} = \frac{(50 \text{ V})}{(152 \text{ }\Omega)} = 0.329 \text{ A}$$

$$i_L = \frac{V}{X_L} = \frac{(50 \text{ V})}{(346 \text{ }\Omega)} = 0.144 \text{ A.}$$

A vector plot would look like Fig. 31-4b.

The maximum current is found from the vector sum to be:

$$I = [(0.5)^2 + (0.329 - 0.144)^2]^{1/2} \text{ A}$$

$$I = 0.533 \text{ A}$$

If we write the voltage as $v(t) = V \sin \omega t$, then the source current is

$$i(t) = (0.533 \text{ A}) \sin (\omega t + \phi)$$

where now

$$\tan \phi = \frac{\left(\dfrac{1}{X_C} - \dfrac{1}{X_L}\right)}{\dfrac{1}{R}}.$$

Please note that this bears no simple resemblance to the phase difference obtained for the series R-L-C circuit.

The time average power in this case is ($\cos \phi = .938$):

$$<P> = (50 \text{ V})(0.533 \text{ A})(1/2)(0.938) = 12.5 \text{ watts}$$

It is instructive to note that the "on resonance" condition is the same for the parallel circuit as it was for the series circuit, namely $X_C = X_L$. The current drawn from a constant voltage source is a minimum here. If we use however a constant current source, then at resonance, the voltage across the parallel combination will be maximum and the power will be maximum, as in the series resonant circuit using a constant voltage source.

Example 7

A transformer connected to a 110 volt line delivers 10 volts to the secondary circuit. If the power drawn from the primary circuit is 220 watts, what is the equivalent resistance R_s of the secondary circuit?

Solution:

Referring to Fig. 31-5, since the flux per turn linked by the primary and secondary circuits is the same, we have the relationship $V_p/N_p = V_s/N_s$.

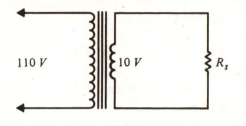

Figure 31-5

and we can solve for the turns ratio

$$\frac{N_p}{N_s} = \frac{V_p}{V_s} = \frac{(110 \text{ V})}{(10 \text{ V})} = 11.$$

Ignoring any power losses in the transformer, we have $V_p I_p = V_s I_s$, so the secondary current is calculable.

$$i_s = \frac{(220 \text{ watts})}{(10 \text{ volts})} = 22 \text{ A}.$$

The equivalent resistance of the secondary is

$$R_s = \frac{V_s}{I_s} = \frac{(10 \text{ V})}{(22 \text{ A})} = 0.455 \ \Omega.$$

The effective resistance in the primary circuit is

$$R_p = \frac{V_p}{I_p} = \frac{(110 \text{ V})}{(2 \text{ A})} = 55 \ \Omega.$$

Note that

$$\frac{R_p}{R_s} = \left(\frac{N_p}{N_s}\right)^2 = (1\,1)^2 = 121.$$

QUIZ

1. A series R-L-C circuit carries current

$$I = (1.25\ A)\ \sin 2\pi(60)t$$

with R = 60 Ω, C = 30 μF, and L = 0.15 H. Calculate the following quantities: (a) the capacitive reactance; (b) the inductive reactance; (c) the impedance; (d) the power factor; (e) the time average power; and (e) the resonant frequency.

Answer: (a) 88.4 Ω; (b) 56.6 Ω; (c) 67.9 Ω; (d) 0.884;(e) 46.9 W; and (f) f_0 = 75.0 s^{-1}.

2. A series circuit has a resistance of 50 Ω and a power factor of 0.80 when the frequency is 80 Hz. The voltage leads the current.
    (a) Calculate the impedance of this circuit.
    (b) Calculate the magnitude of the inductor or capacitor needed to resonate the above circuit assuming the element is placed in series.

Answer: (a) 62.5 Ω; (b) A capacitor C = 53.1 μF placed in series with the above circuit will produce resonance.

3. A 500 Ω resistor, a 5 H inductor, and an 11.1 μF capacitor are connected in *parallel*. Calculate the impedance of this parallel combination when ω is equal to 300 rad/s.

Answer: Z = 300 Ω

4. A lossless transformer with 1000 turns on the primary winding carries a current of magnitude 0.25 A in the primary circuit. If there are 125 turns on the secondary winding, calculate the magnitude of the secondary current.

Answer: I_s = 2.0 A

5. A "lossy" inductor(L) is in series with a "pure" resistor (R) equal to 10 Ω. The voltage across each element is measured and found to be: V_R = (15 Volts) sin [(377 rad/s)t] and V_L = (1.217 Volts) sin [(377 rad/s)t + 1.192 rad] where t is in seconds. Calculate the inductance L and series resistance r of the "lossy" inductor.

Answer: L = 2 mH and r = 0.3 Ω.

32

ELECTROMAGNETIC WAVES

OBJECTIVES

In this chapter, your objectives are to:

Calculate the speed of propagation of electromagnetic waves in vacuum and in various materials.

Calculate the Poynting vector $\vec{S}$ and its time average value given $\vec{E}$ and $\vec{B}$.

Apply the definite relationship between E and B in an electromagnetic wave to calculate their values given the power per unit area (S) in the wave.

REVIEW

Material presented since the introduction to electrostatics is collected and summarized in this chapter. The entire content of all the electromagnetic theory studied previously is contained in four equations known as Maxwell's equations. While it is convenient to incorporate so many diverse phenomena in a small number of equations, the mathematical complexity of these equations is so high that they may not significantly raise the level of understanding of electromagnetic phenomena for every student. A parallel situation occurred when we were able to describe all of classical mechanics in three laws of motion, Newton's laws. Just writing down these laws did not guarantee that we could solve the practical problems met in our study of mechanics.

In this chapter, Maxwell's equations are used to demonstrate the plausibility of wave-like excitations of the electromagnetic field with the following properties:

(1) these waves do not need a material medium for their propagation;
(2) they propagate in a given medium at a fixed and calculable speed;
(3) the waves are transverse waves; the electric and magnetic field vectors are in a plane perpendicular to the direction of propagation;
(4) these waves are capable of transporting energy and momentum.

One of the triumphs of Maxwell's equations is that they permit calculation of the wave velocity from first principles. To obtain this monumental result, the spatial variation of the magnetic field (through the H vector) must be related to the time variation of the electric field. This can be done with Ampere's law when it is modified to include the displacement current that arises from a time changing electric field. Next the spatial variation of the electric field must be related to the time variation of the magnetic field. Faraday's law can be used for this. Then, since the order of taking the derivatives with respect to space and time is immaterial, it is possible to show that both the electric and magnetic fields obey a differential equation that describes wave propagation. From the wave equation it is possible to conclude that the wave speed w satisfies the equation:

$$w^2 = \frac{1}{\mu\varepsilon} = \frac{1}{K_m\mu_0}\frac{1}{K\varepsilon_0}$$

In a perfect vacuum, $K_m = 1$, and $K = 1$ so that the speed there is c, with

$$c^2 = \frac{1}{\mu_0\varepsilon_0}$$

Using the previous values for μ_0 and ε_0, it is found that the speed of electromagnetic waves in a vacuum is 3×10^8 m/s. This is, of course, the speed of light, light being one form of electromagnetic waves.

In the demonstration of the above results, it is found that the electric and magnetic field vectors in the wave are related with $E = cB$. Since $B = \mu_0H$ in vacuum, we find that $E/H = 377$ ohms. This numerical ratio can be very useful in calculations.

The energy density obtained for the combined electric and magnetic fields enables us to examine the energy content of electromagnetic waves. By calculating the product of the energy density and the volume to obtain the energy and then dividing by the cross-sectional area and the time, it is shown that a new vector, $\vec{S}$, the Poynting vector, defined by the equation

$$\vec{S} = \vec{E} \times \vec{H}$$

is equal in magnitude to the power per unit area carried by the wave. The direction of $\vec{S}$ is the direction of wave propagation. One very important example of a Poynting vector is the "solar constant" appropriately labeled S also. The solar constant has a magnitude of about 1.5 kw/m^2 and is a measure of the average power radiated from the sun incident on the earth. A calculation of E and H in sunlight using the known value of the solar constant is given in Example 4. A calculation of the Poynting vector for a wire carrying current is given in Example 5, where the power dissipation per unit length is obtained in two ways.

In addition to transporting energy, electromagnetic waves have momentum, an intriguing concept since they carry no mass. One way of looking at this is to make a generalization of the following nature: the momentum is to be defined as the derivative of the kinetic energy with respect to the speed. This generalization is illustrated in Example 6. Since electromagnetic waves carry momentum and force is equal to the rate of change of momentum, these waves can exert force or give rise to pressures on objects in their path.

As with any wave disturbance, superposition (adding E or B fields) of waves with the same frequency traveling in opposite directions can lead to standing waves. As shown in the text, nodes appear in the pattern for standing waves at distances directly related to the wavelength. Since the maximum distance between nodes is the separation between conducting planes (in the treatment of the text), this leads to a discrete set of frequencies called "normal modes". If the frequency is known, the wave speed can be measured to the accuracy of the product of the distance (between nodes) and the frequency measurement.

QUESTIONS AND ANSWERS

Question. If we label the constant introduced in Coulomb's law, $k_e = 1/4\pi\epsilon_0$, and that constant introduced when calculating the magnetic field, $k_m = \mu_0/4\pi$. What is the value of the ratio k_e/k_m and what is its significance?

Answer. The value of the ratio is $(9 \times 10^9)/(10^{-7})$ SI units $= 9 \times 10^{16}$ (m/s)^2. This ratio is the square of the speed of light.

Question. If electromagnetic waves, such as light, can exert pressure, would absorbed light or reflected light exert more force on a surface for normal incidence?

Answer. The reflected light would exert twice as much force as absorbed light since the momentum change would be twice as large.

EXAMPLES AND SOLUTIONS

Example 1

A light wave with frequency 5×10^{14} Hz is incident from air on a material with dielectric constant 7.5 and relative permeability 1.25. The frequency of the light wave is unchanged as it passes from air into the other medium. Calculate the wavelength of the light wave in air and in the material.

Solution:

The dielectric constant and relative permeability of air are so close to unity that we will take the speed of light in air to be the same as the speed of light in vacuum, namely $c = 3.00 \times 10^8$ m/s. The wavelength in air is then

$$\lambda_a = \frac{c}{f} = \frac{\left(3 \times 10^8 \text{ m/s}\right)}{\left(5 \times 10^{14} \text{ /s}\right)}$$

$$= 6 \times 10^{-7} \text{ m} = 600 \text{ nm}$$

In the other material, the wave speed w is

$$w = \frac{1}{\sqrt{\epsilon\mu}}$$

$$w = \frac{c}{\sqrt{K K_m}}$$

$$w = \frac{(3 \times 10^8 \text{ m/s})}{\sqrt{(7.5)(1.25)}} = 0.980 \times 10^8 \text{ m/s}.$$

The corresponding wavelength is

$$\lambda_m = \frac{w}{f} = \frac{(0.980 \times 10^8 \text{ m/s})}{(5 \times 10^{14} \text{ /s})}$$

$$= 1.96 \times 10^{-7} \text{ m} = 196 \text{ nm}.$$

Example 2

A plane electromagnetic wave is propagating in air (free space) in the $+ x$ direction with

$$\vec{H}(x, \ t) = H_0 \sin 2\pi \left(ft - \frac{x}{\lambda} \right)(\hat{\jmath})$$

Calculate (a) $\vec{E}(x,t)$ and (b) the instantaneous value of the Poynting vector, $\vec{S}$. In this expression for the wave, f is the frequency and λ is the wavelength.

Solution:

(a) $\vec{E}$ must point in the z direction since $\vec{E}$ is perpendicular to $\vec{B}$ (or $\vec{H}$) and they both are perpendicular to the direction of propagation. As the direction of propagation is given by the direction of $\vec{S}$,

$$\vec{S} = \vec{E} \times \vec{H}$$

then **E** must be

$$\vec{E}(x, \ t) = E_0 \sin 2\pi \left(ft - \frac{x}{\lambda} \right)(- \hat{k})$$

The amplitude E_0 is calculable from H_0 as

$$E_0 = (377 \ \Omega) H_0$$

The final form is

$$\vec{E}(x, \ t) = - (377 \ \Omega) \sin 2\pi \left(ft - \frac{x}{\lambda} \right)(\hat{k})$$

(b) The Poynting vector $\vec{S}$ is given by

$$\vec{S} = \vec{E} \times \vec{H} = (-\hat{k} \times \hat{j})\left(\frac{H_0^2}{wK\varepsilon_0}\right) \sin^2 2\pi\left(ft - \frac{x}{\lambda}\right)$$

Thus S points in the + x direction and its time average value, <S>, is equal to:

$$\langle S \rangle = \frac{1}{2}\left(\frac{H_0^2}{wK\varepsilon_0}\right)$$

Example 3

A source of electromagnetic waves with power 10^7 radiates uniformly in all directions. Calculate the amplitude of the electric field vector for these waves
      (a) at a distance of 100 m from the source
      (b) at a distance of 1 km from the source.

Solution:

The number given for the power represents the time average power of the source. To find the power per unit area at a distance R from the isotropic source construct a spherical surface of radius R about the source. The surface area is then

$$A = 4\pi R^2$$

The power per unit area for R = 100 m is then

$$\frac{P}{A} = \frac{10^7 \text{ W}}{4\pi \times 10^4 \text{ m}^2} = 79.6 \text{ W/m}^2$$

This power per unit area is equal to the time average value of the Poynting vector, <S>:

$$\langle S \rangle = \frac{1}{2} EH = 79.6 \text{ W/m}^2$$

Using

$$(E/H) = 377 \ \Omega$$

to replace H, we can solve for E

$$\frac{1}{2}\frac{E^2}{(377 \ \Omega)} = 79.6 \text{ W/m}^2$$

$$E^2 = 6.00 \times 10^4 \ (\text{W}\cdot\Omega \ /\text{m}^2)$$

E = 245 V/m

(b) At a distance of 1 km from the source, the power per unit area is

$$S = \frac{P}{A} = \frac{10^7 \text{ W}}{4\pi \times 10^6 \text{ m}^2} = 0.796 \text{ W/m}^2$$

Solving for E as above, we have

$$\frac{1}{2}\frac{E^2}{(377 \ \Omega)} = 0.796 \text{ W/m}^2$$

$$E^2 = 6.00 \times 10^2 \ (\text{W}\cdot\Omega \ /\text{m}^2)$$

$$E = 24.5 \text{ V/m}$$

The amplitude E is proportional to R^{-1} where R is the distance from the source.

Example 4

The time average power from the sun falling on the earth can be represented by the solar constant, S, which is approximately equal to 1.4 kW/m². What are the maximum values of E and H due to this electromagnetic radiation?

Solution:

The Poynting vector magnitude, when $\vec{E}$ and $\vec{H}$ are perpendicular, is equal to EH. If E and H are sinusoidal functions of time, the time average power, according to Eq. 32-24 is:

$$\langle S \rangle = \frac{1}{2} EH$$

To simplify the numerical calculation, recall that E/H = 377 Ω yielding

$$\langle S \rangle = \frac{1}{2} E \left(\frac{H}{E}\right)\left(\frac{E}{H}\right) H = \frac{1}{2} E \left(\frac{H}{E}\right)(377 \ \Omega) H$$

$$\langle S \rangle = \frac{1}{2} (377 \ \Omega) H^2$$

Solving numerically we have

$$H = \sqrt{\frac{(2 \times 1.4 \times 10^3)}{377}} = 2.73 \text{ A/m}$$

$E = (377\ \Omega)H = 1.03 \times 10^3$ V/m.

Example 5

A wire of radius a, cross sectional area A, and resistivity ρ carries a current I.

    (a) Find the magnetic intensity, H, at the surface of this wire.

    (b) Find the electric field strength from Ohm's law.

    (c) Calculate the Poynting vector magnitude and direction.

    (d) Calculate the power crossing the surface area of a cylinder of height L and radius a using the Poynting vector and show that it is equal to the usual expression for power dissipation in a conductor.

Solution:

(a) From Ampere's law we have that the field, B, at the surface of the wire is:

$$B \cdot 2\pi a = \mu_0 I$$

Since $B = \mu_0 H$ when there is no magnetization, we have $H = I/2\pi a$, with a direction given by the right hand rule as shown in Fig. 32-1.

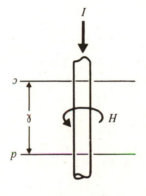

Figure 32-1

(b) The potential difference between c and d, V_{cd} is given by:

$$\int_c^d \mathbf{E} \cdot d\mathbf{L} = I \left(\text{resistance between c and d}\right)$$

For this uniform wire, E is constant so we have

$$EL = I\rho\ L/A \text{ so } E = \rho(I/A)$$

The direction of E is the same as the direction of I here.

(c) Since E points along the wire and H is tangent, $\vec{E} \times \vec{H}$ points radially inward at all points on the surface and the magnitude of $\vec{E} \times \vec{H}$ is just EH.

$$S = EH = \rho \frac{I}{A} \frac{I}{2\pi a}$$

(d) The power (P) crossing the surface for the wire of length L is given by multiplying the constant value of S by the side area of the cylinder of height L and radius a, $2\pi a L$.

$$P = 2\pi a S = EH = \rho \frac{L}{A} I^2$$

Since $\rho L/A$ is the resistance of this length of wire, it is seen that the power calculated this way is also equal to $I^2 R$ but here it is taken from the electromagnetic field and flows radially inward!

Example 6

(a) Generalize the definition of momentum to make momentum equal to the derivative of the kinetic energy with respect to speed and then calculate the momentum Δp in an infinitesimal volume of space.
(b) Using Newton's laws, find the radiation pressure, P.

Solution:

(a) Consider the infinitesimal volume ΔV to have cross-sectional area A and length $c\Delta t$ where c is the speed of light. The energy density, u, is given by:

$$u = \frac{1}{2} \varepsilon_0 E^2 + \frac{1}{2} \mu_0 H^2 = \varepsilon_0 E^2$$

Regarding this as kinetic energy, we have for the kinetic energy in the volume ΔV

$$\Delta E_K = \varepsilon_0 E A c \Delta t$$

Let's check the new definition of momentum by returning to classical mechanics. There,

$$E_K = \frac{1}{2} mv^2 \quad \text{so,} \quad \frac{d}{dv}(E_K) = mv$$

which was just our previous result. Try this new definition on the kinetic energy in the radiation field:

$$\Delta p = \frac{d}{dc}(\Delta E_K) = \frac{d}{dc}\left(\varepsilon_0 E^2 A c \Delta t\right)$$

Thus we have

$$\Delta p = \varepsilon_0 E A \Delta t$$

which is the momentum contained in the volume element.

(b) From Newton's law we write the force, F, as $F = \Delta p/\Delta t$.

$$F = \frac{\Delta p}{\Delta t} = \left(\varepsilon_0 E^2 A\right)$$

The pressure, P, is the force per unit area (F/A) so we have for the pressure

$$P = \varepsilon_0 E = u, \text{ the energy density in the field.}$$

Example 7

A capacitance C has plates that are circular with area $A = \pi R^2$. The two plates are separated by distance d as shown in Figure 33-2a. An alternating voltage source $v(t) = V \sin \omega t$ is connected across the plates as shown creating a current $I(t)$ in the external circuit and a "displacement" current $I_D(t)$ between the plates. Find the magnetic field produced in this case.

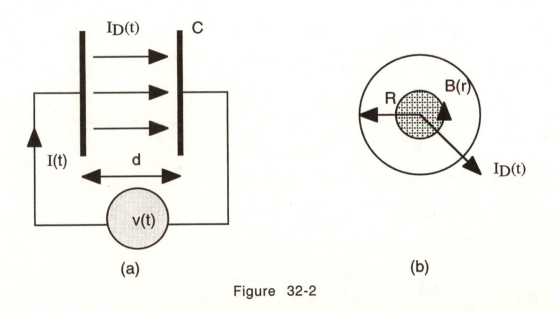

Figure 32-2

Solution

The left capacitor plate is shown in Figure 32-2b, with the direction of the displacement current pointing up (or out of the page). For this direction of $I_D(t)$, the magnetic field lines will be counter-clockwise as shown. First we calculate the displacement current so that the current is continuous in the circuit and between the plates.

$$I_D(t) = \frac{dQ}{dt} = \frac{d(Cv)}{dt} = C\frac{dv}{dt} = \omega CV \cos \omega t$$

This current, uniformly distributed in space over the surface of the circular plates can be characterized by a current per unit area J with $I_D(t) = JA$. To find B in the region between the plates and outside the plates, we use Ampere's law:

$$\int_C \vec{B} \cdot d\vec{s} = \mu_0 (I_{true} + I_D)_{enc}$$

In this case, between the plates there is no "true" current so only the displacement current contributes to B. Referring to Figure 32-2b, the vector B has the same magnitude for every point on the circle of radius r (shaded region) and the direction is always tangent to the circle so that the "dot" product is the same at each point on the circle. Therefore the left-hand side of the integral becomes:

$$\int_C \vec{B} \cdot d\vec{s} = B \int_C ds = 2\pi r B$$

The right-hand side of the Ampere's law relationship is:

$$\mu_0 (I_{true} + I_D)_{enc} = \mu_0 (0 + JA_{enc}) = \mu_0 (J\pi r^2) \quad \text{for } r \leq R$$
$$\mu_0 (I_{true} + I_D)_{enc} = \mu_0 (I_D) \quad \text{for } r \geq R$$

Equating the results for the two sides, we find:

$$2\pi r B(r) = \mu_0 (J\pi r^2) \quad \text{for } r \leq R$$
$$2\pi r B(r) = \mu_0 (I_D) \quad \text{for } r \geq R$$

Simplifying these expressions and substituting for J;

$$B(r,t) = \frac{\mu_0}{4\pi} \left(\frac{2\omega CV}{R^2} r \cos \omega t \right) \quad \text{for } r \leq R$$

$$B(r,t) = \frac{\mu_0}{4\pi} \left(\frac{2\omega CV}{r} \cos \omega t \right) \quad \text{for } r \leq R$$

Inside the plates, B is linearly proportional to r but outside the plates it decreases as (1/r). Note that the direction of the instantaneous Poynting vector S is radially inward. Would the time average of S be equal to zero?

192

QUIZ

1. An electromagnetic wave is characterized by an electric field vector

$$\vec{E}(z, t) = (1.6 \text{ V/m}) \sin 2\pi \left(ft + \frac{z}{\lambda} \right) (\hat{\imath})$$

Find the wave form that describes the magnetic intensity H.

Answer:

$$\vec{H}(z, t) = (4.24 \times 10^{-3} \text{ A/m}) \sin 2\pi \left(ft + \frac{z}{\lambda} \right) (-\hat{\jmath})$$

2. At a distance of 15 km from a source that radiates uniformly in all directions, the electric field amplitude is found to be 125 V/m. Calculate
        (a) the magnetic intensity amplitude H,
        (b) the time average value of the Poynting vector,
        (c) the time average power radiated by the source.

Answer: (a) H = 0.332 A/m
             (b) $\overline{S}$ = 20.7 W/m^2
             (c) $\overline{P}$ = 5.86 x 10^{10} W.

3. The antenna of a radio station radiates equally in all directions. The total power of the transmitter is 50,000 W. The power per unit area is equal to the magnitude of the Poynting vector, S. Calculate the value of S, in watts per square meter, at a distance of 100 km from the station.

Answer: S = 3.99 x 10^{-7} W/m^2

4. In a plane wave propagating in the +x direction, the Poynting vector has the magnitude 1.508 x 10^{-3} W/m^2. The magnetic intensity vector (H) has a magnitude of 2 x 10^{-3} S.I. units and points in the +y direction. Give the magnitude and direction of the electric field vector, E, for this plane wave.

Answer: 0.754   V/m in the -z direction

5. Solar "furnaces" (concentrators of sunlight) are being designed that will have the intensity of "50,000 Suns". Calculate the electric field strength for a source having power 50,000 times the solar constant S (1.4 kW/m^2).

Answer: E = 2.3 x 10^5 V/m.

33

THE NATURE AND PROPAGATION OF LIGHT

OBJECTIVES

In this chapter your objectives are to:

Identify that part of the electromagnetic spectrum that stimulates the retina of the eye as light.

Describe light waves as transverse waves with E and B fields perpendicular to each other and to the direction of propagation.

Review the historic experiments that measured the velocity of light.

Formulate the laws of reflection and refraction.

Apply the laws of reflection and refraction to a variety of problems such as total internal reflection.

Identify several of the different methods of producing polarized light waves, such as absorption, reflection, scattering, and birefringence.

Distinguish between linear, circular, and elliptical polarizations and make calculations concerned with devices such as quarter wave plates.

Calculate the reduction in intensity obtained when polarizing plates are stacked with various relative orientations.

REVIEW

Whether light should be thought of as a wave-like disturbance or a stream of particles has been a topic of considerable interest for nearly 400 years. Sir Isaac Newton believed that light consisted of particles and explained many known optical phenomena based on his theory. Unfortunately his theory required that the speed of light in a material medium be larger than the speed in vacuum. As we saw in the last chapter, the reverse is true: the speed of light is highest in a vacuum. (The space between the earth and the sun is not a perfect vacuum but it is a pretty good one.) While for the most part, light behaves as a wave, the particle-like nature of light is observed in atomic, nuclear, and high energy physics.

Because of its very large magnitude (c = 3 x 10^8 m/s), the speed of light proved difficult to measure on the earth by ordinary means. Roemer, an astronomer, made the first measurement with any reasonable accuracy but he did not actually claim that he had measured c. It was not until 1850 that a value for c was obtained using terrestrial measurements. This method, due to Fizeau, is illustrated in Example 1.

Since "light" is that portion of the electromagnetic spectrum that affects the retina of the eye, it is important to know what part of the spectrum we are dealing with. Our maximum sensitivity occurs at about a wavelength of 550 nm (where a nanometer is 10^{-9} m) with most eyes being reasonably sensitive from 400 nm (violet) to 700 nm (red).

Reflection and refraction of waves are generally easier to picture and understand if we use rays to indicate the direction in which the waves are moving. These rays are perpendicular to the actual wave front (or locus of equal phase). In using such rays is should be noted that:

(1) the incident ray, the reflected ray, the transmitted (refracted) ray and the normal to the interface between two surfaces all lie in the same plane;

(2) the angle of incidence is equal to the angle of reflection (all angles are measured with respect to the normal);

(3) Refraction of rays obeys Snell's law, $n_a \sin \phi_a = n_b \sin \phi_b$, where n_a and n_b are respectively the indices of refraction (measured with respect to vacuum) of media a and b, and ϕ_a, ϕ_b are the angles between the rays and the normal. The index of refraction is a function of the frequency or wavelength. Unless otherwise specified, we will ignore this effect and use an average value.

Snell's law predicts an interesting phenomena known as "total internal reflection". For this to occur, light must be traveling from a more dense (higher index of refraction) to a less dense medium (lower index of refraction) as in going from water to air. In this case, there exists a critical angle of incidence ϕ_c for which the refracted ray will be bent parallel to the interface ($\phi = 90°$) and not emerge into the less dense medium. For angles of incidence greater than the critical angle, the rays are totally reflected back into the dense medium. If we designate medium 'a' as the dense medium, the critical angle for total internal reflection is a solution of the equation:

$$\left(\sin \phi_a\right)_c = \frac{n_b}{n_a}$$

This is an essential result for the currently interesting field of fiber optics. Of additional interest in this connection is the absorption of light by matter, even high quality optical glass. It is shown in the text that the light intensity decreases exponentially with distance into the material. The absorption coefficient is a material dependent parameter and must be made very small for fiber optic bundles of any significant length, like telephone lines.

By using every point on a wave front as a source of secondary wavelets, Huygens found that he could predict the wave front at a later time from the known front at an earlier time. Using the Huygens' construction, it is possible to show that:

$$\frac{\sin \phi_a}{\sin \phi_b} = \frac{v_a}{v_b}$$

where v_a and v_b are the propagation velocities in media "a" and "b". Since $n_a \sin \phi_a = n_b \sin \phi_b$, then $n_a v_a = n_b v_b$ and we can relate the index of refraction to the propagation velocity. In particular if we choose medium "a" to be vacuum (free space) so that $n = 1$ and $v = c$, then $n_b = c/v_b$. Since $v_b \leq c$ then $n_b \geq 1$. When a wave passes from one medium to another, while the velocity of propagation does change, the frequency does not. Thus $\lambda_a/v_a = \lambda_b/v_b$ or stated in terms of n's, $\lambda_a n_a = \lambda_b n_b$.

Waves of all wavelengths propagate in the vacuum with the same speed but in a material medium, waves with different wavelengths travel with different speeds. The medium is said to be dispersive or to exhibit dispersion. There is no dispersion in free space. This dispersion can be easily observed by looking at the fan of colors that emerges from a prism illuminated with white light.

The phenomenon of polarization exists only for transverse waves such as electromagnetic waves, waves on a string, etc., but not for longitudinal waves such as sound waves. When a charged particle accelerates, it radiates electromagnetic waves. The electric field vector is proportional to this acceleration vector and parallel to it. Knowledge of the direction of propagation and the electric field vector determines the magnetic field vector. By convention, the direction of polarization is the direction of the electric field vector. Light emitted from a typical source is unpolarized, even though the light emitted in an individual transition from a single charge is polarized. The acceleration vectors of the various charges in the source are randomly oriented so the net effect is an unpolarized beam.

To make the discussion of polarization more quantitative the text introduces the concept of an ideal polarizing filter that passes all light polarized in the direction of the filter's axis. Thus if unpolarized light is incident on such a filter and the resultant light projected on a screen, the intensity of light on the screen is less than it would be if the filter was removed but independent of the orientation of the filter axis. This is due to the fact that the electric field vector has the same magnitude on average for any direction in a plane perpendicular to the direction of propagation for unpolarized light. After the light has passed through this first filter, it is *linearly polarized* in that its electric field vector now points along the filter axis. It is not correct to think that all other vectors were rejected as the transmitted light would be very weak in that case. Rather it is better to think of each E vector not aligned with the filter axis as being resolved into components parallel and perpendicular to this axis. The parallel components are passed but the perpendicular ones are rejected. The resultant intensity after passing unpolarized light through one ideal filter is just one half the initial intensity of the unpolarized beam. Then if the E vector of polarized light makes an angle of θ with respect to the filter axis, only the component, E cos θ, is passed through the filter. Since the intensity, I, is proportional to the square of the electric field vector, the transmitted intensity is then:

$$I = I_{max} \cos^2 \theta,$$

where I_{max} is the intensity transmitted when $\theta = 0°$ (but is only one half of the original unpolarized source intensity). See Example 9.

A second filter whose axis is perpendicular to that of the first filter will now pass none of the polarized light incident on it. This is clear from the above expression for the intensity as now we have made $\theta = 90°$ so the cosine vanishes. It is quite interesting that under these conditions of *complete extinction*, introduction of a third filter in between the first two crossed filters will cause light to be transmitted unless its axis accidentally coincides with that of one of the other filters. This point is illustrated in Example 10.

Another method of producing polarized light is by reflection. The explanation of this phenomenon is contained in a detailed solution of Maxwell's equations at an interface. We will only state without proof that it happens. Furthermore, when the angle between the reflected beam and the refracted ray is exactly 90°, the reflected beam will be perfectly linearly polarized. This fact enables one to derive Brewster's law, namely

$$\tan \phi_p = \frac{n'}{n}$$

where n' is the refractive index of the medium containing the incident and reflected rays and n is the refractive index of the medium containing the refracted ray. Light incident at the angle ϕ_p will be totally polarized upon reflection, with the direction of polarization perpendicular to the plane containing the incident, reflected and refracted rays. This effect is examined in Examples 11 and 12.

Polarization of light by double refraction is also possible. This means that the material exhibits two different indices of refraction depending on the direction of the E vector. For this to occur the refracting medium (crystal) must have highly anisotropic properties. If the direction of propagation is taken as the z axis, such a crystal basically sorts the incoming unpolarized wave and sends the component of $\vec{E}$ along the y axis on a different path from the x component. Such a crystal can be used to produce linearly polarized light by blocking out one ray or it can produce circularly polarized light by recombining the two beams after they have gone through a specific distance. See Example 13 for an illustration of this point.

Light scattered from small particles suspended in in a liquid (or in air) is linearly polarized when viewed at right angles to the incident light. To understand this effect, think of the scattered light as radiation that was absorbed from the incident beam and then re-radiated. Since light cannot be polarized in the direction of propagation, the scattered light viewed at 90° with respect to the incident beam is linearly polarized with its direction of polarization perpendicular to the plane containing the incident and scattered beams. The geometry pertinent to this effect is given in Example 14.

HINTS AND PROBLEM-SOLVING STRATEGIES

The main text gives problem-solving strategies for the problems in this chapter. Now is a good time to review those hints. As in our study of mechanics, a sketch is of vital importance for these problems. The student is also reminded that all angles are to be measured from the normal to a surface, not the surface itself. For problems dealing with the polarization of light, the direction of polarization is that of the electric field vector.

QUESTIONS AND ANSWERS

Question. Is it possible for a material to have more than one refractive index?

Answer. Yes. First, the refractive index can depend on the wavelength of light (dispersion) and secondly, it can depend on the state of polarization (birefringence).

Question. Many painters, some of the great observers of nature, have created rainbows in their compositions that have the colors inverted from actual rainbows. What is the correct order of the colors? Is red on the top or bottom?

Answer. In the "primary" rainbow, formed by rays that are internally reflected once in the water drops, blue (purple) is on the bottom, yellow is in the middle, and red is at the top. The "secondary" rainbow is fainter and requires two internal reflections. How are its colors arranged?

EXAMPLES AND SOLUTIONS

Example 1

In the experiment performed by Fizeau using a toothed wheel, suppose the distance between the wheel and the mirror is 10 km. Calculate the angular velocity needed so that a light ray passing through the center of one tooth and then reflected by the mirror just passes through the center of the next tooth. There are 60 teeth in the rotating wheel.

Solution:

We must calculate two travel times and then equate them. First we calculate the time for the light beam to pass through tooth one and then return to the rotating wheel. The total distance is 20 km so $t = 20 \times 10^3$ m/c. The angular distance between centers of teeth is $360°/60 = 6°$. Converting this to radians, we have the angle the wheel must turn through as $\theta = \pi/30$ rad. Since $\theta = \omega t$ for constant angular velocity, then $t = \theta/\omega$. Thus

$$\frac{\theta}{\omega} = \frac{20 \times 10^3 \text{ m}}{c}$$

Solving for ω,

$$\omega = \frac{\left(3 \times 10^8 \text{ m/s}\right)\left(\pi/30\right)}{\left(20 \times 10^3 \text{ m}\right)} = 500\pi \text{ rad/s}$$

This corresponds to a frequency of 250 Hz. Note that at all integral multiples of this frequency, we would still get an image of our light source.

Example 2

For the light ray shown in Fig. 33-1, the angle of incidence is 60°. The same ray on leaving the flat glass on the other side is displaced a distance $d = 0.80 \times 10^{-2}$ m from the spot where it entered the glass. Calculate the index of refraction for the glass plate of thickness 1.2×10^{-2} m.

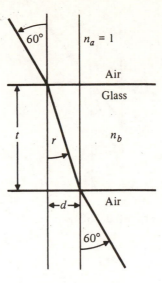

Figure 33-1

Solution:

Using the main text's problem-solving strategy, note the the angles in Figure 33-1, a sketch of the problem situation, are measured with respect to the normals. From Snell's law,

$$n_a \sin 60° = n_b \sin r$$

Here $n_a = 1$, so $n_b = \sin 60°/\sin r$. From Fig. 33-1, $\tan r = d/t$ and both d and t are given so r can be calculated. Since

$$\sin r = d/(d^2 + t^2)^{1/2} = 0.5547,$$

$$n_b = \frac{0.8660}{0.5547} = 1.561$$

Example 3

A coin rests on the bottom of a shallow pond of depth 1 m. Taking the index of refraction of water to be n = 4/3, find the "apparent depth" of the coin when viewed normal to the air-water interface. See Fig. 33-2.

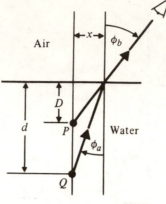

Figure 33-2

Solution:

To obtain the answer for the above question, we must solve another problem first. A ray coming from the coin is refracted at the surface as shown, but the eye traces the ray back along a straight line and one thinks the ray originated at point P (apparent depth D), not at point Q (depth d). Using Snell's law, $\sin \phi_b = n_a \sin \phi_a$, taking $n_b = 1$. From the geometry of the figure, we have

$$\sin \phi_a = \frac{x}{\sqrt{x^2 + d^2}} \quad \text{and} \quad \sin \phi_b = \frac{x}{\sqrt{x^2 + D^2}}$$

Thus

$$\frac{n_a x}{\sqrt{x^2 + d^2}} = \frac{x}{\sqrt{x^2 + D^2}} .$$

By squaring both sides we obtain

$$n_a^2(x^2 + D^2) = x^2 + d^2.$$

This is the desired equation that relates the true depth (d) to the apparent depth (D). To apply this equation to "normal" viewing (from above) we take the limit as x goes to zero obtaining $n_a D = d$. Numerically we have

$$D = \frac{d}{n_a} = \frac{3}{4} \text{ m.}$$

Example 4

A point light source 2 m below the surface of water produces a circular pattern of light when viewed from above. Taking the index of refraction of water to be n = 4/3, calculate the radius of this circle.

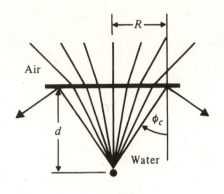

Figure 33-3

Solution:

Light rays from the point source which approach the water-air interface with an angle of incidence less than the critical angle will be partially transmitted and partially reflected so we will see light from them. Those rays approaching at an angle of incidence equal to or greater than the critical angle will be totally internally reflected. Since there is symmetry in this problem about the perpendicular line from the point source to the interface, the pattern seen from above will be a circle (not uniformly illuminated). Referring to Fig. 33-3, the critical angle ϕ_c is a solution of the equation $n_a \sin \phi_a = n_b \sin \phi_b$ with $\sin \phi_a$ set equal to 1 if $n_a < n_b$. In this case $n_a = 1$ so $(\sin \phi_b)_c = 1/n_b$. From the geometry of the figure

$$(\sin \phi_b)_c = R/(R^2 + d^2)^{1/2}$$

so we equate the two expressions for $(\sin \phi_b)_c$ to obtain:

$$\frac{R}{\sqrt{R^2 + d^2}} = \frac{1}{n_b} \quad \text{or} \quad \frac{Rn_b}{\sqrt{R^2 + d^2}} = 1$$

Solving this for the unknown R, since $(Rn_b)^2 = R^2 + d^2$, we have

$$\frac{R}{d} = \frac{1}{\sqrt{n_b^2 - 1}} = 1.134 \quad \text{or} \quad R = 2.27 \text{ m}.$$

Example 5

Suppose a camera held underwater is pointed straight up toward the water-air interface (a fish eye view). What part of the horizon will appear in a photo taken in this manner?

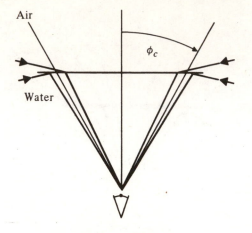

Figure 33-4

Solution:

Referring to Fig. 33-4, some rays near the critical ray for total internal reflection are shown. The critical angle $\phi_c = \sin^{-1}(3/4)$ if we take n for water to be 4/3. This angle is $d_c = 48.59°$. Thus *the entire horizon will be visible* but the hemispherical image will be contained in a distorted way in a cone of angle ϕ_c as shown in Fig. 33-4.

Example 6

Suppose a flat glass plate ($n_b = 1.561$) rests on a layer of water (n = 4/3). What angle of incidence in air (n = 1) will just give total internal reflection at the glass-water surface?

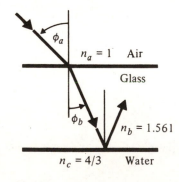

Figure 33-5

Solution:

Referring to Fig. 33-5, the angle of incidence at the glass-water interface is ϕ_b, the angle of refraction at the air-glass interface. For total reflection, we have

$$\sin \phi_b = \frac{n_c}{n_b}$$

Also, $\sin \phi_a = n_b \sin \phi_b$, so $\sin \phi_a = n_c$ for total reflection. This would make the sine greater than unity so *this condition is impossible.* We can conclude that there is no angle of incidence in air that will produce total internal reflection at the glass-water interface.

Example 7

For the light ray incident on the prism as shown in Fig. 33-6, calculate the difference in angle between the path of the emerging ray and the incident ray. For A = 60°, n = 1.50, and an angle of incidence of 30°, how large is the deviation?

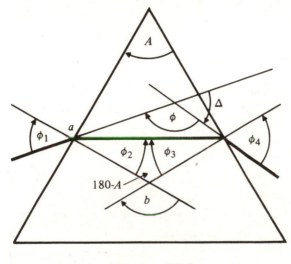

Figure 33-6

Solution:

If the prism angle is A, the normals intersect at an angle of 180° - A. For the polygon abcd the sum of angles must be 360° so we have

$$\phi_1 + \phi_4 + 180° - A + \Delta = 360°$$

where $\phi + \Delta = 180°$. Δ is the desired deviation in the directions of the emerging beams. Thus

$$\phi_1 + \phi_4 = A + \Delta.$$

Also we have

$$\phi_2 - \phi_3 + 180° - A = 180° \text{ or } \phi_2 + \phi_3 = A.$$

Snell's law gives the following relations:

$$\sin \phi_1 = n \sin \phi_2$$

$$n \sin \phi_3 = \sin \phi_4$$

Given n, A, and ϕ_1, we can find ϕ_2, ϕ_3, and ϕ_4 and then Δ. In this case A = 60° and ϕ_1 = 30°. Numerically we have

$$1.5 \ \sin(60° - \phi_3) = 0.5 \quad \text{so } \phi_3 = 40.53°$$

$$1.5 \ \sin(40.53°) = \sin \phi_4 \quad \text{so } \phi_4 = 77.10°$$

Thus

$$\Delta = 30° + 77.10° - 60° = 47.10°$$

Since a prism like this disperses the spectrum, measurements of Δ for the various colors allow you to determine the index of refraction for the various wavelengths (at least roughly).

Example 8

For the prism geometry studied in the previous example, let the prism angle A be small and let ϕ_1, ϕ_2, ϕ_3, and ϕ_4 be small enough so that the small angle approximation can be used for Snell's law. Show that the angular deviation, Δ, is independent of the angle of incidence ϕ_1 (an important result for the theory of thin lenses).

Solution:

Summarizing the important results:

(a) $\phi_1 + \phi_4 = A + \Delta$

(b) $\phi_2 + \phi_3 = A$ (so if ϕ_2 and ϕ_3 are small, so is A)

(c) $\sin \phi_1 = n \sin \phi_2$

(d) $n \sin \phi_3 = \sin \phi_4$

Using the small angle approximation for (c) and (d) gives $\phi_1 = n\phi_2$ and $n\phi_3 = \phi_4$. If we add these two equations,

$$\phi_1 + \phi_4 = n\phi_2 + n\phi_3 = n(\phi_2 + \phi_3).$$

From (b) we conclude that $\phi_1 + \phi_4 = nA$ so substituting back into (a) we have:

$$nA = A + \Delta.$$

Thus

$$\Delta = (n - 1)A.$$

The angle through which a ray is bent is thus the same for all rays in this approximation.

Example 9

A beam of unpolarized light with intensity I_0 is incident on a perfect polarizing filter. How is the intensity of the transmitted beam, I, related to I_0?

Solution:

Using the main text's problem-solving strategy, we note that an unpolarized beam is a random mixture of all polarization states and that the direction of polarization is that of the electric field. The unpolarized beam can be thought of as a collection of electric field vectors, E, that make all possible angles, θ_i, with respect to any given axis (such as that of the filter). The transmitted component will be $E \cos \theta_i$ and the corresponding intensity will be proportional to $E^2 \cos^2\theta_i$ where we must take the average value of $\cos^2 \theta_i$. Since any value of θ_i between 0 and 2π is equally likely, this average value is 1/2. Thus $I = 1/2\ I_0$. This is a physically reasonable result indicating that the ideal filter absorbs half the unpolarized energy falling on it.

Example 10

Two ideal polarizing filters are arranged so that the light intensity passing the second filter is zero (i.e. their axes are perpendicular). A third such filter is introduced *between the first two*. What is the total transmitted light intensity as a function of the orientation of this third filter?

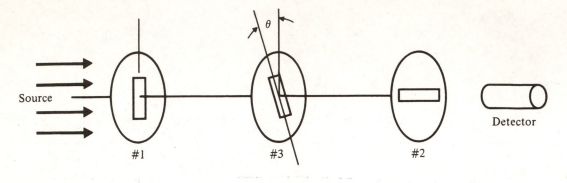

Figure 33-7

Solution:

Refer to Fig. 33-7. Let θ be the angle between the axes of filters 1 and 3. If E_0 is the magnitude of the electric field vector that passes through filter 1, then the component passed through filter 3 is $E_0 \cos \theta$. The component of this vector passed through filter 2 is now

$$(E_0 \cos \theta) \cos(90° - \theta) = E_0 \cos \theta \sin \theta.$$

Using the main text's problem-solving strategy, we note that the intensity is proportional to the square of the electric field amplitude. The intensity at the detector is then:

$$I \propto E_0^2 \cos^2 \theta \sin^2 \theta$$

Since the maximum transmitted intensity (all axes in same direction) is proportional to E_0^2, we can write:

$$I = I_{max} \cos^2 \theta \sin^2 \theta$$

$$I = \frac{I_{max}}{4} \sin^2(2\theta)$$

This obviously has its maximum when $\theta = 45°$. The third filter has served to rotate the direction of polarization.

Example 11

What is the relationship between Brewster's angle, θ_p, and the critical angle for total internal reflection?

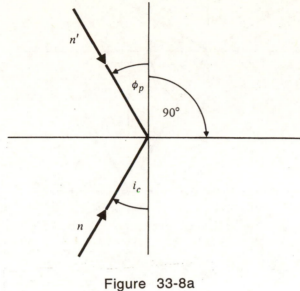

Figure 33-8a

Solution:

For critical internal reflection to be possible at the interface shown in Fig. 33-8a, n must be greater than n' and the critical angle i_c is obtained from:

$$n \sin i_c = n'$$

For the reflected beam in the medium with index n', totally polarized light results when the angle between the reflected and refracted beams is 90°. Thus

$$n' \sin \phi_p = n \sin \phi = n \sin(90° - \phi_p) = n \cos \phi_p.$$

Dividing the first expression by the last yields:

$$\tan \phi_p = (n/n')$$

Notice that it is not necessary for n' < n for this effect. However if n and n' are such that critical internal reflection is possible, then

$$(\tan \phi_p)(\sin i_c) = 1.$$

In Fig. 33-8b, note that if the angle of incidence is equal to ϕ_p so that the reflected beam, R', is totally polarized, then the beam R_2 is also totally polarized as it makes a 90° angle with the

207

transmitted beam I' at the second interface.

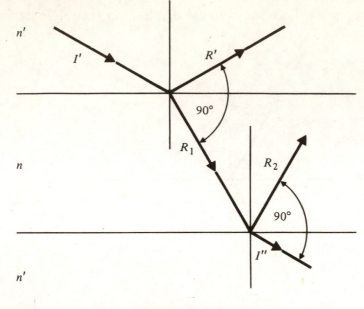

Figure 33-8b

Example 12

In the previous example
        (a) suppose n' = 1.00 and n = 1.50 (glass)
        (b) suppose n' = 1.00 and n = 4/3 (water)
        (c) suppose n' = 1.50 and n = 4/3 (water)
and calculate the Brewster angle ϕ_p.

Solution:

This involves only a simple substitution in the equation

$$\tan \phi_p = (n/n')$$

(a) $\tan \phi_p = 1.50$ so $\phi_p = 56.3°$

(b) $\tan \phi_p = 4/3$ so $\phi_p = 53.1°$

(c) $\tan \phi_p = 8/9$ so $\phi_p = 41.6°$

Thus the reflection-polarization can occur at the interface of any two optical media. The ratio n/n' determines whether the Brewster angle is greater or less than 45°.

Example 13

Consider two waves of the same frequency $\omega = 2\pi c/\lambda$ traveling in the $+ z$ direction with amplitudes A_1 and A_2 given by,

$$\vec{E}_1 = \hat{\imath}A_1 \sin(kz - \omega t)$$

$$\vec{E}_2 = \hat{\jmath}A_2 \sin(kz - \omega t + \phi)$$

so that ϕ gives the relative phase between E_1 and E_2. Specify A_1, A_2, ϕ for
   (a) a linearly polarized wave,
   (b) a circularly polarized wave, and
   (c) an elliptically polarized wave.

Solution:

(a) If $\phi = 0$ we have a linearly polarized wave, where $E_1/A_1 = E_2/A_2$ so that a plot of E_1 versus E_2 is a straight line through the origin. If $A_1 = A_2$, the direction of polarization is at 45° with respect to our arbitrarily chosen x and y axes so that by rotating the coordinates to a new system x', y' we can have a wave polarized along x' or one polarized along y'.

(b) If $\phi = \pm \pi/2$ we have circularly polarized light if $A_1 = A_2$. In this case $E_1^2 + E_2^2 = A_1^2$ so that a plot of E_1 versus E_2 gives a circle of radius A_1.

(c) If $\phi = \pm \pi/2$ but $A_1 \neq A_2$ we have elliptically polarized light since $E_1/A_1 = E_2/A_2 = 1$. A plot of E_1 versus E_2 is an ellipse. If the two waves of light are recombined in a quarter wave plate, the emerging single beam in circularly polarized.

Example 14

A beam of light (initially unpolarized) is passed through an ideal polarizing filter P_1 and then traverses a cell containing small particles suspended in water as shown in Fig. 33-9. Light scattered from these small particles is observed at right angles to the incident beam after passing it through a second ideal filter P_2. Originally the filters are oriented so that the "brightness of the field" as seen by an observer is maximum.

     (a) Filter P_2 is rotated through 90°. What does the observer see?

     (b) Filter P_1 is now rotated through 90°. Is the field bright or dark?

     (c) Filter P_2 is now rotated to its original position. Is the field bright or dark?

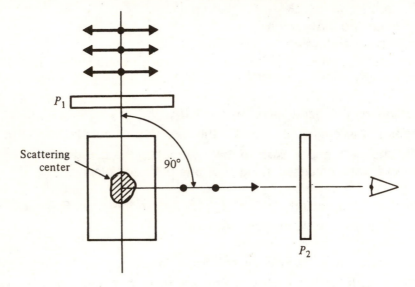

Figure 33-9

Solution:

     This is like a detective story. To have maximum brightness initially, P_2 must transmit waves with E vectors perpendicular to the plane of the paper because the scattered beam at 90° contains only those components. Clearly P_1 must transmit those waves as well, or the scattered light intensity at 90° could not be maximum.

(a) If P_2 is rotated through 90°, the observer will see a dark field.

(b) If P_1 is rotated through 90°, it absorbs all the waves with E vectors perpendicular to the plane of the paper. The field remains dark.

(c) If P_2 is now rotated back to its original position, the field remains dark because P_1 is not passing any waves with E vectors perpendicular to the plane of the paper.

QUIZ

1. The angle of incidence in medium 1 is 45° and the angle of refraction in medium 2 is 30°. The index of refraction of medium 2 is 1.510. Calculate the index of refraction in medium 1.

Answer: n_1 = 1.068

2. A liquid of index of refraction n floats on top of water of index of refraction n_w = 1.33. Light rays in water that make an angle of incidence of 60° or larger are totally reflected. Calculate n.

Answer: n = 1.152

3. Unpolarized light of intensity I_0 strikes a combination of three ideal polarizing filters. The first and third polarizers are "crossed" so that their axes are perpendicular. The intensity of light transmitted by this combination is 0.1152 I_0. Calculate the angle θ between the axes of the first and third filters.

Answer: θ = 36.9° or 53.1°

4. *Choose the correct statement.*
   Light polarized by scattering from small particles in suspension in a liquid
   (a) is completely polarized only when viewed along the original beam
   (b) is never completely polarized
   (c) is completely polarized when viewed at 90° with respect to the incident wave, with the electric field vector parallel to the incident beam
   (d) is completely polarized when viewed at 90° with respect to the incident wave, with the electric field vector perpendicular to the plane of the incident wave and the scattered wave
   (e) none of these.

Answer: Statement (d) is correct.

5. A light ray in air is incident at an angle of 30° with respect to the normal on side 1 of a sheet of plastic with refractive index of n = 1.45. How thick should the slab be in order for the ray emerging from side 2 to be displaced 3 cm from the incident ray?

Answer: t = 8.17 cm.

34
GEOMETRIC OPTICS and OPTICAL INSTRUMENTS

OBJECTIVES

In this chapter, the images formed by reflection or refraction at a single surface (either plane or spherical) are studied. These results are then applied to study images formed by two or more refracting surfaces. Your objectives are to:

Locate the image of a single surface by ray construction (graphical methods).

Calculate the position of the image using general expressions developed from geometry and Snell's law; characterize the image as erect or inverted; and find its relative size.

Apply sign conventions for object and image distances as well as those for the radius of curvature of the surface.

Calculate the focal length of a lens made from a pair of refracting spherical surfaces.

Locate the images formed by two or more surfaces graphically by ray tracing.

Calculate the image distance and the lateral and longitudinal magnification given the object distance and the focal length.

Apply simple lens formulas to calculate the properties of optical instruments such as magnifiers, cameras, and telescopes.

REVIEW

In this chapter, reflection and refraction at a single surface are studied in preparation for a treatment of lenses and more complicated optical instruments that involve multiple surfaces. A single sign convention, suitable for both reflection and refraction, is introduced. Since only plane and spherical surfaces are considered (with results for the plane surface derivable from those for the spherical surface by letting the radius approach infinity), only signs for three distances will be needed. These are (1) the object distance, s; (2) the image distance, s'; and (3) the radius of curvature, R. All of these distances are measured from the intersection of the reflecting or refracting surface with the optic axis, a reference line along which we position the object and image. The sign conventions are the following:

(a) The object distance (s) is positive if it is on the same side of the surface as the incoming light;

(b) The image distance (s') is positive if it is on the same side of the surface as the outgoing light;

(c) The radius of curvature (R) is positive if the center of curvature, C, is on the same side of the surface as the outgoing light. This is the same convention as that used for s'.

If there is only one reflecting or refracting surface to deal with, all of the object distances, s, are positive, basically by construction. In the more complicated systems, image formation depends on treating the image formed by surface 1 as the object for surface 2, etc., until all surfaces have been treated. Here the object distance for surfaces 2, 3, etc. are frequently negative. Simple numerical examples of both concave (positive R) and convex (negative R) spherical mirrors are found in Examples 1 and 2. The general relationship between object distance and image distance is shown graphically in Example 3. The results for small distances (s) may seem a little perplexing at first but at such small distances (s << R), the curvature of the surface is unimportant and the results closely resemble those for a plane (flat) mirror. Note that a concave mirror can produce an image that is either magnified or reduced and either inverted or erect, but the convex mirror can produce only a reduced, erect image of a real object.

There are two ways of defining a focal length or focal point. In one method, the object is imagined to be located at infinity and the corresponding image distance is called the second focal length (f'). In the other method, one seeks the position of the object that would give an image at infinity. The object distance here is called the first focal length (f). For spherical mirrors f = f' = R/2 where R is positive for a concave mirror but negative for a convex mirror. For a single spherical refracting surface, f is not equal to f'.

Two types of magnification are introduced, a lateral magnification (m) which is equal to the ratio of the vertical height of the image (y') to that of the object and a longitudinal magnification (m') which is equal to the ratio of the differential change in image position (s') to a corresponding differential change in object position (s). The values of m and m' for all the surfaces encountered in this chapter are summarized in the text in Table 34-1. Negative values for m always indicate that the image is inverted with respect to the object whereas a positive m accompanies an erect image.

All derivations in this chapter make use of the small angle (or paraxial) approximation at some stage. Only rays near the optic axis can be used (hence the small angles) as the geometry of the spherical reflecting (refracting) surface is such that the various rays that can be constructed to locate the image do not all intersect in the *same point* for an object of finite size. This imperfect imaging is called "spherical aberration".

It is desirable to construct a "ray diagram" for each optics problem involving lenses, mirrors or combinations of these. For spherical surfaces, a ray through C (the center of curvature) is always undeviated as the angle of incidence is zero. Rays parallel to the optic axis are either reflected (or refracted) through a focal point or they are reflected (or refracted) in such a way that they appear to originate from a focal point. The rules for forming principal rays are summarized for mirrors in the text.

Ray tracing to locate the image for a concave spherical refracting surface is illustrated in Example 4. There also the general expression developed in the text for such a surface is applied to calculate the image position and the lateral magnification. Use of this general formula to treat a plane refracting surface is illustrated in Example 5 where the plane surface is realized by taking the limit as R approaches infinity.

Finally, a derivation of the formula for refraction from a spherical surface (concave surface with negative radius of curvature) is given in Example 6. Only plane geometry and no special sign conventions are used. This is done to emphasize that all we are doing in this chapter is applying plane geometry, Snell's law and the law of reflection in the small angle approximation to locate the image due to a given object.

The basic result relating the image and object distances to the respective indices of refraction and the radius of curvature of a spherical refracting surface, is used to obtain the focal length of a thin lens consisting of two such spherical refracting surfaces. The procedure employed is to find the image due to the first surface and then use that image as the object for the second surface. The image formed by the second surface is the "final" image for this thin lens. If the focal length is found from the lensmaker's formula (or known otherwise), the object distance (s), image distance (s') and focal length (f) are related (in air or free space) by the formula

$$\frac{1}{s} + \frac{1}{s'} = \frac{1}{f}$$

and the two spherical refracting surfaces that make up the lens are thus treated as a whole.

Lenses with a positive value of f are known as converging lenses while those with negative values of f are called diverging lenses. Two definitions of the focal point are given and these are illustrated in Example 7 where the principal rays used to locate the image graphically are also shown for both converging and diverging lenses.

The lateral magnification, m, defined as the ratio of image height to object height, is shown to be equal to -s'/s. If m is a negative number, the image is inverted but if m is positive, the image is erect. The longitudinal magnification, m', is shown to be equal to $-m^2$. Applications of these general ideas to image formation by converging and diverging lenses are given in Examples 8 and 9. The general relationship between s and s' for both converging and diverging lenses is shown graphically in Example 10.

Two thin lenses in contact with each other can be treated as we did the two spherical refracting surfaces in order to derive the lensmaker's formula. This is pursued in Example 11 where it is shown that if the individual lenses have focal lengths of f_1 and f_2, the two together, considered as a single system, have a focal length, f, found from:

$$\frac{1}{f} = \frac{1}{f_1} + \frac{1}{f_2}$$

Optical instruments frequently have more than one lens or mirror for their function. To locate the image of a given object for a complicated system, the image of the given object due to the first reflecting or refracting surface encountered by the light is found by standard methods. This image is then treated as the object for the next surface. The "object" distance must be measured from the second surface and obey the previously stated sign convention. The image from this surface is then treated as the object for the next surface, etc.

Six applications of geometric optics are described: the camera, the projector, the eye, the magnifier, the microscope, and the telescope. Each application is treated using the thin lens approximation of geometric optics although commercial designs of these objects use more sophisticated approaches.

THE CAMERA. The camera uses a single lens mounted in a light tight enclosure in its simplest version. Film is used for recording the *real image* produced by the lens. The distance between the film and the lens can be varied to permit imaging distant objects as well as nearby objects. In practice the single lens is actually replaced by a lens combination. The "travel" of a camera lens relative to the film is treated in Example 12.

For a camera, the field stop limits the size of the exposed film to a rectangle typically of dimensions 24 mm x 36 mm. An aperture stop of diameter D controls the amount of light that enters the camera. The "f-stop" of a camera is the ratio of the focal length, f, to the diameter of the aperture stop, D.

$$f \text{-} stop = \frac{f}{D}$$

The amount of light that strikes the film is proportional to the intensity of the light reflected (or emitted) by the source, the inverse square of the distance between source and lens (s^{-2}) and the area of the aperture ($\pi D^2/4$). See Example 13.

THE PROJECTOR. The slide projector consists of two lenses (a condensing lens and a projection lens) and a light source. The condensing lens is actually a pair of lenses that insures that the light striking the outer portions of the slide also strikes the projection lens--giving the projected slide a more uniform appearance. The focal length of the condensing lens is chosen to make the image of the source slightly smaller than the usable diameter of the projection lens.

The projection lens must have a diameter that is larger than the diagonal of a rectangular slide. The focal length of this lens is chosen to give the appropriate magnification. For a fixed distance between the slide and the screen, the distance between the projection lens and the slide is varied until the image of the slide is in sharp focus on the screen. See Example 14.

THE EYE. The human eye is a very sophisticated optical instrument. It is nearly spherical in shape with a diameter of about 2.3 cm. Its major components are: the *retina*, which plays the role of the film in a camera; the *cornea*, a tough transparent material covering the eye, responsible for most of the refraction; the *lens* (a converging lens) with an index of refraction about 1.44; the *iris* a muscular diaphragm that controls the size of the *pupil*; and the pupil which regulates the quantity of light that reaches the retina.

Frequently the power of a lens is expressed in a unit called diopters. The power (P), in diopters, is the inverse of the focal length (f) *expressed in meters*.

$$P = \frac{1}{f}$$ where f is expressed in meters.

The cornea typically has a power of 40 to 45 diopters while the lens itself has only a power of about 20 to 24 diopters. For thin lenses in contact, the total power of the lens system is the sum of the individual powers, so the eye has an overall power (as a lens) of between 60 and 65 diopters.

The relaxed *normal* eye produces sharp images of distant objects on the retina. As objects are brought closer to the eye, since the image distance is fixed, the focal length of the eye must decrease. This process is called accommodation and results from the muscles altering the shape of the lens. The *near point* of the eye is the closest distance of an object from the eye for which a sharp image can be formed. We will adopt a value of 25 cm for the typical near point distance as we discuss other optical instruments and assume that the eye cannot focus on objects that are closer than 25 cm from the lens. This distance is also used in the definition of angular magnification (M). Examples 15 and 16 consider lenses to correct vision problems.

THE MAGNIFIER. The simple magnifier employs just one converging lens and is used to produce an enlarged image of a nearby object. Since the eye is used to view this image the final image is on the retina of the eye.

It is conventional to treat the magnifier as though the image is at infinity, so that it is viewed by the "relaxed eye" and is less tiring to view. The angle subtended by the image at the eye is then:

$$\theta_\infty \cong \frac{h}{f}$$

Thus the angular magnification (M) is

$$M = \frac{\theta_\infty}{\theta_M} \cong \frac{h/f}{h/25 \text{ cm}} = \frac{25 \text{ cm}}{f} \ .$$

where θ_M is the angle subtended at the eye when the object is viewed (without the magnifier) at the near point. This arrangement produces a smaller magnification than an arrangement where the final image is produced at the near point of the eye as shown in Example 17. Since typical values of the angular magnification, M, lie between 2 and 4, the difference between forming the image at ∞ or at the near point can be significant.

THE MICROSCOPE. When it is desirable to observe greater detail than is possible with the simple magnifier, two lenses can be used to increase the magnification. Such an arrangement is called a compound microscope or simply a microscope. The object O to be viewed is placed just outside the focal point of lens 1 (called the objective lens) and a real inverted image I_1 is produced. This image then serves as the object for the second lens (called the eyepiece) which acts like the simple magnifier described in the last section. The total magnification is then the product of the lateral magnification of the objective and the angular magnification of the eyepiece.

In practice, f_o and f_e, the focal lengths of the objective and eyepiece respectively, are of the order of a centimeter. The length of the tube, L, that houses the lenses is of the order of fifteen centimeters. Treating the lenses as thin lenses for simplicity, we can arrive at an approximate formula for the magnification. Applying the thin lens result to the lateral magnification of the objective,

$$m = -\frac{s'}{s}$$

If the image I_1 falls just at the focal point of the eyepiece, then the "relaxed eye" (placed very close to this lens) views the final image (a real, inverted image) I at infinity. The angular magnification is thus M_∞ or

$$M_\infty = \frac{25\ cm}{f_e}$$

The distance s' is equal to $L-f_e$ which is approximately L and the distance s is approximately equal to f_o. The total magnification, with this approximate treatment, is

$$M_{total} \cong -\left(\frac{L}{f_o}\right)\left(\frac{25\ cm}{f_e}\right)$$

The negative sign indicates that the final image I is inverted with respect to the object. Magnifications of the order of 400 are not difficult to achieve with inexpensive lenses. Example 18 treats the compound microscope.

THE TELESCOPE. Another example of a two lens (in principle) system is the telescope. The arrangement is very similar to that of the microscope except the microscope is intended to view small objects that are placed relatively close to the objective. The microscope produces an image larger than the original object. On the other hand, the astronomical telescope is intended to view large, distant objects and produce angular magnification of the object. The actual size of the image is usually much smaller than the actual size of the object.

The objective lens of the astronomical telescope serves to gather the light from the distance object (shown as parallel rays in the figure) and produce a sharp inverted image (h') at its focal point. For viewing the image with the relaxed eye, the image is placed at the focal point of the eyepiece--producing the final image at infinity--still inverted with respect to the object. The angular magnification of the eyepiece is the ratio θ_e/θ_o since θ_o is the angle subtended at the unaided eye by the distant object.

$$M = -\left(\frac{h'/f_e}{h'/f_o}\right) = -\left(\frac{f_o}{f_e}\right).$$

(the angles are evaluated on the side of the lens where light is incident, so $\theta_0 = h'/fo$ as the "object" is erect on that side but $\theta_e = -h'/fe$ since the image is inverted.) From this expression is is apparent that large angular magnification is possible if f_o is large and f_e is small.

Another variation of the astronomical telescope is called the Galilean telescope. Here a diverging lens is used for the eyepiece so that the final image is erect. Otherwise, the Galilean telescope is the same as the conventional telescope. For terrestrial applications, it is desirable that the final image be erect. Another method for producing an erect final image is to insert a third converging lens (with focal length f') between the objective lens and eyepiece. In Example 19, it is shown that if this lens is chosen so that the final magnification of the telescope is the same, the "terrestrial" telescope is longer than the astronomical telescope by a distance 4f'.

HINTS AND PROBLEM-SOLVING STRATEGIES

The main text gives problem-solving strategies for both mirrors and simple lenses (and systems of simple lenses). We will refer to them in the problems. Now is a good time to review those strategies. It is extremely important to note that the principal ray diagrams made in this chapter are as important as free-body diagrams in our study of mechanics.

Use the previously discussed procedures for finding the final image of an optical system. With light starting at the left in the sketch of the problem situation, work through the system, one refracting (reflecting) surface at a time. The image of the first surface becomes the object for the second and so forth.

QUESTIONS AND ANSWERS

Question. In side view mirrors on automobiles "objects are closer than they appear". Is the mirror concave or convex.

Answer. Convex. Referring to Fig. 34-2, the convex mirror produces a virtual image which is smaller and hence the object (reduced in size) seems further away than it really is.

Question. A reflecting telescope's mirror has a radius of curvature of 50 cm. Explain how to use the telescope to observe a solar eclipse.

Answer. Place a screen 25 cm from the mirror where a real image of the sun will appear. Do not look into the mirror at the sun. Look at the screen.

Question. A person is nearsighted. What kind of lens (diverging or converging) is needed to correct the person's vision?

Answer. A near sighted person sees nearby objects (diverging rays) clearly. So the distant, parallel rays must be made to diverge so that the object appears closer. Needs a diverging lens.

Question. The near sighted person takes off his glasses and attempts to use them as magnifying glasses. Are they a success?

Answer. No. The diverging lens produces a virtual image of a nearby object that appears smaller than the object.

EXAMPLES AND SOLUTIONS

Example 1

A concave spherical mirror with R = 0.8 m produces an image of an object located 2 m from the mirror. Where is the image located and what is its relative size?

Solution:

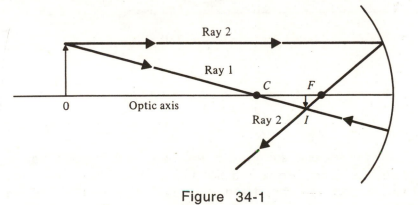

Figure 34-1

Following the problem-solving strategy of the main text, a principal ray diagram is given in Figure 34-1. Ray 1 passes through the center of curvature and is undeviated. Ray 2, parallel to the axis, is reflected through the focal point F located at R/2. The intersection of these two rays locates the image, I. Using the formula for a mirror,

$$\frac{1}{s} + \frac{1}{s'} = \frac{2}{R}$$

we note that s, the object distance, is positive since it is on the same side of the mirror as incoming light. The image distance, s', will also be positive since the image is on the same side of the mirror as outgoing (or reflected) light. Furthermore, since C is on the same side of the mirror as outgoing light, R is positive. Numerically we have

$$\frac{1}{s'} = \frac{2}{(0.8 \text{ m})} - \frac{1}{(2 \text{ m})} \qquad \text{or } s' = 0.50 \text{ m}$$

The image is located graphically on the principal ray diagram and its position (and size) agree with the above calculation. Since the lateral magnification, m, is equal to -s'/s, then m = -1/4. The image height is only 1/4 of the object height and the minus sign indicates that the image is inverted.

Example 2

A convex spherical mirror with R = 0.8 m produces an image of an object 2 m from the mirror. Where is the image located and what is its relative size?

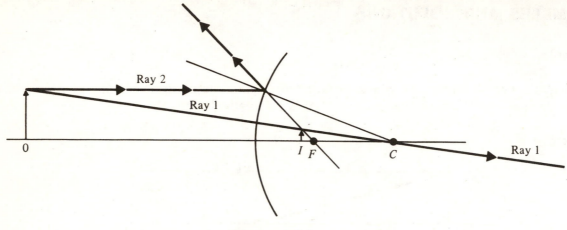

Figure 34-2

Solution:

Referring to the principal ray diagram given in Fig. 34-2, we see again that Ray 1 passes through C and is undeviated. Ray 2 is reflected and appears to originate at F. The small, erect image I is shown at the intersection of rays 1 and 2. Since C is not on the same side of the mirror as outgoing light, R is negative. From the figure, since the image is also not on the same side as outgoing light, we expect s' to be negative. Numerically we have

$$\frac{1}{s} + \frac{1}{s'} = \frac{2}{R} \quad \text{gives} \quad \frac{1}{2\text{ m}} + \frac{1}{s'} = \frac{2}{-\ (0.8\text{ m})}$$

Solving for s' yields

$$s' = -\ (1/3)\ \text{m}$$

Note the negative sign and that location on the sketch agrees with the calculation. The lateral magnification is m = − s'/s = + 1/6, the (+) sign indicating that the image is erect.

Example 3

Since both concave and convex spherical mirrors obey the same equation, construct a graph of object distance versus image distance for each type of mirror.

Solution:

(a) Concave mirror.

$$\frac{1}{s} + \frac{1}{s'} = \frac{2}{R}$$

where R is positive. See Fig. 34-3a.

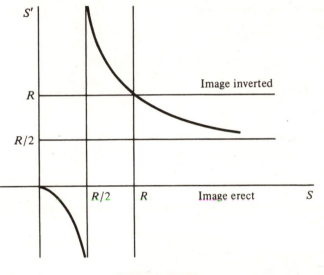

Figure 34-3a

(b) Convex mirror. If we let R stand for a positive number, |R|, then

$$\frac{1}{s} + \frac{1}{s'} = -\frac{2}{|R|}$$

See Fig. 34-3b.

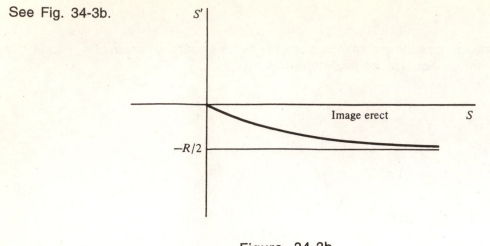

Figure 34-3b

Example 4

A concave spherical (|R| = 0.8 m) refracting surface (n' = 1.50) forms an image of an object placed 1.6 m away in air. Locate the image and calculate the lateral magnification.

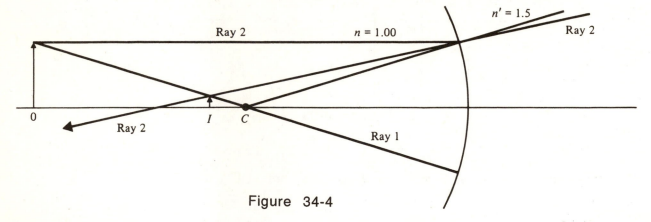

Figure 34-4

Solution:

Refer to Fig. 34-4 where the ray diagram is shown with ray 1 passing through C and ray 2 refracted according to Snell's law. The intersection of ray 1 with the extension of ray 2 locates the image, I. The general formula from the text,

$$\frac{n}{s} + \frac{n'}{s'} = \frac{(n' - n)}{R}$$

222

is used here with s = 1.6 m, n = 1.00, n' = 1.50, and R = - 0.8 m. R is negative here because C is *not* on the same side of the surface as the outgoing light. Thus we have

$$\frac{1.5}{s'} = \frac{(1.5 - 1.0)}{(-0.8 \text{ m})} - \frac{1.0}{1.6 \text{ m}}$$

Solving for s' gives s' = 1.2 m. Again, from the principal ray diagram we see that the image location agrees with the calculation. The lateral magnification, m, is equal to – ns'/sn' so m = + 0.50, the positive sign again indicating that the image is erect. Note that the distance IC is 0.4 m and OC is 0.8 m. Since the triangles are similar, we have y'/y = 0.4/0.8 = 0.5 again.

Example 5

Use the expression developed for refraction at a spherical surface to obtain the corresponding result for a plane refracting surface. For the spherical surface note that

$$\frac{n}{s} + \frac{n'}{s'} = \frac{(n' - n)}{R}$$

where n is the index of refraction of the medium where the object is located, n' is the index of refraction of the other medium and R is the radius of curvature of the surface.

Solution:

For a plane surface, the radius of curvature is ∞, so we have

$$\frac{n}{s} + \frac{n'}{s'} = 0$$

Thus for a plane surface the result is

$$\frac{s'}{s} = -\frac{n'}{n}$$

as given in the text.

It is important to observe the correct sign conventions here. If s is positive, then s' is always negative but it can be either larger or smaller than s depending on the ratio n'/n. To apply this formula to find the "apparent depth" of an object in water when we view it in air, regard s' as the apparent depth, s as the true depth, n' = 1 and n = 4/3. See Example 3, Chapter 36. The negative sign occurs because the image is not on the same side of the surface as the *outgoing* light.

Example 6

For an object as shown in Fig. 34-5, locate the image due to refraction at the spherical surface. Assume n'< n for convenience. Calculate both the longitudinal and lateral magnifications.

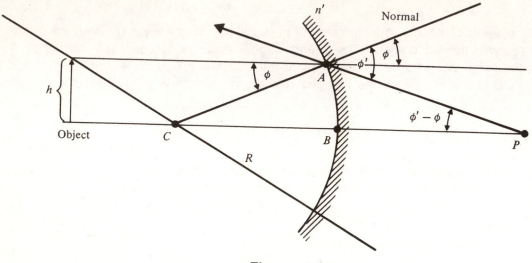

Figure 34-5

Solution:

This problem can be easily solved by using plane geometry and Snell's law. We will use an xy coordinate system with origin at point B. The ray from the tip of the object to A is refracted at the surface and passes through P. Designate the height of the object by h. An equation that describes the straight line through points P and A is

$$y - h = - \tan(\phi' - \phi) x$$

The ray passing from the tip of the object through the center of curvature, C, is undeviated and an equation describing that line is:

$$y = - \left(\frac{h}{s - R} \right)(x + R)$$

where s is the distance from the object to B so that h/(s - R) is the slope of the line through C.

If we eliminate y from these two equations, we obtain the value of x for which these two lines intersect. This is where the image is formed, o x = s' in the notation of the text. To evaluate tan(ϕ' - ϕ) we can use Snell's law and the small angle approximation,

$$n \sin \phi = n' \sin \phi' \quad \text{or} \quad n\phi = n'\phi'$$

to find that

$$\tan(\phi' - \phi) = \phi' - \phi$$

is approximately equal to

$$\left(\frac{n}{n'} - 1\right)$$

The value of ϕ here is h/R, so we have

$$\tan(\phi' - \phi) = \frac{h}{R}\left(\frac{n - n'}{n}\right)$$

Solving for x we have

$$x\, Rn - s(n - n') = -\, Rn's$$

Rearranging this result,

$$\frac{Rn - s(n - n')}{Rs} = -\frac{n'}{x}$$

yields

$$\frac{n}{s} + \frac{n'}{x} = \frac{(n - n')}{R}$$

This is the same as the general formula in the text if we set x = s' and remember that R is negative for this particular surface.

As a specific application, take n' = 1 and position the object at the center of curvature (s = R). Then the result is

$$\frac{n}{R} + \frac{1}{x} = \frac{n}{R} - \frac{1}{R}$$

so

$$x = -\, R.$$

The image of this object is located just at the position of the object. The lateral magnification in this case is:

$$m = \frac{y}{h}$$

where y is found by substituting x = - R into the equation

$$y - h = -(x)\tan(\phi' - \phi).$$

225

Here we find that y = nh so m = n. The image is erect and n times larger (vertically) than the object.

To obtain the longitudinal magnification, return to the expression

$$\frac{n}{s} + \frac{n'}{s'} = \frac{(n' - n)}{R}$$

and calculate ds'/ds. Note that

$$\frac{n}{s^2} ds + \frac{n'}{(s')^2} ds' = 0 \quad \text{so} \quad \frac{ds'}{ds} = -\frac{n}{n'} \frac{(s')^2}{s^2}$$

Since s'² = s² in this application and n' = 1, the longitudinal magnification is equal to the lateral magnification. This is not generally true.

Example 7

Sketch the three principal rays for both a converging (f > 0) and diverging (f < 0) lens and show the first and second focal points.

Solution:

Refer to main text's problem-solving strategies for thin lenses. In this example, we methodically provide the principal ray diagrams.

(a) The object point for which the image is at infinity is called the first focal point (F). See Fig. 34-6a for its principal ray.

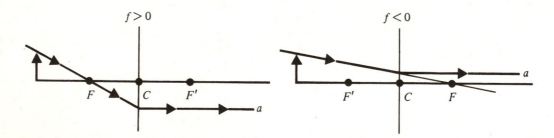

Figure 34-6a

226

(b) The image point for an infinitely distant object is called the second focal point (F'). See Fig. 34-6b for its principal ray.

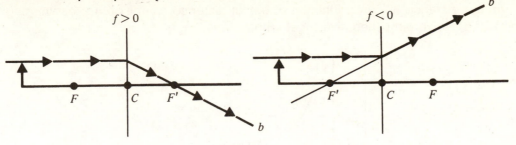

Figure 34-6b

Thus F and F' are located on opposite sides for converging and diverging lenses. For a given *thin* lens, F and F' are numerically the same.

(c) The ray through the center of a thin lens is deviated by a negligible amount (which we take to be zero) and hence is the same for either a converging or diverging lens. See Fig. 34-6c for this principal ray.

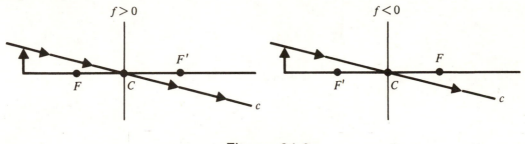

Figure 34-6c

(d) We combine these three principal rays to locate the image due to a real object. The rays are labeled a, b and c, following the previous diagrams. The intersection of these rays locates the tip of the image (I). See Fig. 34-6d.

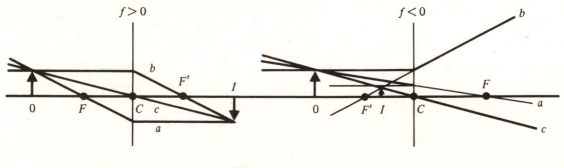

Figure 34-6d

Note: Rays a and b must be projected back to locate I.

227

Example 8

Graph the image distance (s') as a function of object distance (s) for both a converging (positive focal length) and a diverging (negative focal length) lens.

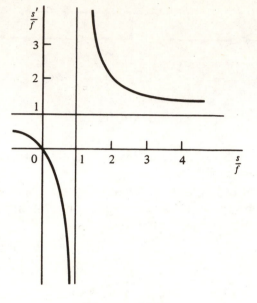

Figure 34-7a

Solution:

(a) For a converging lens, see Fig. 34-7a, we plot the equation

$$\frac{1}{s} + \frac{1}{s'} = \frac{1}{f}$$

Note: To make a "universal" graph of such a function, multiply by f so that

$$\frac{f}{s} + \frac{f}{s'} = 1$$

and then measure s and s' in units of f.

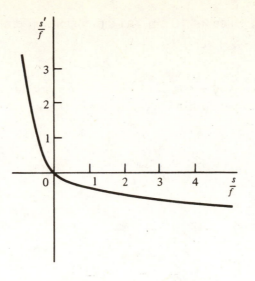

Figure 34-7b

(b) For a diverging lens, see Fig. 34-7b, we plot the equation

$$\frac{|f|}{s} + \frac{|f|}{s'} = -1$$

Example 9

For glass with n = 1.50, use the lensmaker's formula to find the focal lengths for lenses with:

      (a) R_1 = 5 cm, R_2 = -20 cm
      (b) R_1 = 5 cm, R_2 = ∞
      (c) R_1 = 5 cm, R_2 = 10 cm

After finding the focal lengths, calculate the image position and lateral magnification for an object placed + 10 cm from the lens.

Solution:

The lensmaker's formula is

$$\frac{1}{f} = (n - 1)\left(\frac{1}{R_1} - \frac{1}{R_2}\right)$$

(a) R_1 = 5 cm and R_2 = - 20 cm, so for f in centimeters,

$$\frac{1}{f} = (0.5)\left(\frac{1}{5} + \frac{1}{20}\right) = \frac{1}{8} \quad \text{or} \quad f = 8 \text{ cm.}$$

229

To locate the image due to an object at 10 cm, (all dimensions in cm)

$$\frac{1}{10} + \frac{1}{s'} = \frac{1}{8} \quad \text{or} \quad s' = +40 \text{ cm.}$$

The lateral magnification is m = y'/y = s'/s = -4 here. The negative sign indicates the image is inverted. A sketch of this lens would look like Fig. 34-8.

Radius 20 cm

Figure 34-8

(b) R_1 = 5 cm and R_2 = ∞ (this means that the back surface is flat) and thus we have

$$\frac{1}{f} = (0.5)\left(\frac{1}{5} - 0\right) \quad \text{so f} = +10 \text{ cm.}$$

In this case, the object is located at the focal point so the image is at infinity and the magnification is infinite.

(c) R_1 = 5 cm and R_2 = 10 cm (both are positive so the lens is crescent shaped)

$$\frac{1}{f} = (0.5)\left(\frac{1}{5} - \frac{1}{10}\right) = \frac{1}{20} \quad \text{so f} = 20 \text{ cm.}$$

To locate the image write:

$$\frac{1}{10} + \frac{1}{s'} = \frac{1}{20} \quad \text{so s'} = -20 \text{ cm.}$$

This image is located on the same side of the lens as the object. The lateral magnification is m = s'/s = +2, so the image is enlarged and erect (+ sign).

Example 10

For glass with n = 1.50, use the lensmaker's formula to find the focal lengths for lenses with:

      (a) $R_1 = -5$ cm, $R_2 = 20$ cm

      (b) $R_1 = -5$ cm, $R_2 = \infty$

      (c) $R_1 = -5$ cm, $R_2 = -10$ cm

Calculate the image positions and lateral magnifications for an object placed + 10 cm from each of the above lenses.

Solution:

The values of all the R's above are identical to those of the previous example but the signs have all been changed to convert those converging lenses into diverging lenses. Thus each value for f will have the same magnitude but the signs will be negative here.

(a) $f = -8$ cm; $\dfrac{1}{s'} = -\dfrac{1}{8} - \dfrac{1}{10}$

Solving for s' gives s' = - 40/9 cm and m = + 4/9, and the image is reduced and erect. The image is located on the same side of the lens as the object.

(b) f = - 10 cm. Here the object is placed at the wrong focal point to produce an image at infinity and the image distance is found from

$$\frac{1}{s'} = -\frac{1}{10} - \frac{1}{10} = -\frac{1}{5}$$

or s = - 5 cm. Thus m = + 0.5 so the image is on the same side of the lens as the object, erect and reduced in size.

(c) $f = -20$ cm; $\dfrac{1}{s'} = -\dfrac{1}{20} - \dfrac{1}{10} = -\dfrac{3}{20}$

Solving for s' gives s' = - 20/3 cm so that m = + 2/3. Again the image is on the same side of the lens as the object, reduced and erect. Consult the graph of s' versus s in Example 2 for a diverging lens and you will see that the image is always on the same side as the object (negative s'), erect (negative s') and reduced (- s' < s).

Example 11

Suppose two thin lenses with focal lengths f_1 and f_2 are placed in contact with each other. Obtain an expression for the focal length f of this lens combination. {Note that this example illustrates the problem-solving strategy in the main text where the image for the first part of the problem becomes the object for the second part.}

Solution:

This problem is a good example of the systematic method used to solve optics problems when more than one reflecting or refracting surface is involved.

For lens (1) locate an object at s_1. The image distance s_1' is calculable from:

$$\frac{1}{s_1'} = \frac{1}{f_1} - \frac{1}{s_1}$$

This image of lens 1 is now regarded as the object for the second lens. Note that if s_1' is a positive number in the above expression, the image would be on the same side of the lens combination as outgoing light from lens 1. Thus as an object for lens 2, it would give a *negative* object distance (s_2). If however s_1' were negative this would correspond to a positive object distance for the second lens. Thus $s_2 = -s_1'$ is correct for both possibilities, and we have

$$\frac{1}{s_2} + \frac{1}{s_2'} = \frac{1}{f_2}$$

becomes

$$-\left(\frac{1}{f_1} - \frac{1}{s_1}\right) + \frac{1}{s_2'} = \frac{1}{f_2}$$

Rearranging

$$\frac{1}{s_1} + \frac{1}{s_2'} = \frac{1}{f_1} + \frac{1}{f_2}$$

Interpreting this expression, we note s_1 is the distance of the object from the lens combination, and s_2' is the distance of the image from the lens combination, so the focal length of the combination (f) satisfies the equation

$$\frac{1}{f} = \frac{1}{f_1} + \frac{1}{f_2}$$

where now we write s_1 as simply s (the object distance for the combination) and s_2' as s' (the image distance for the combination). Thus we have

$$\frac{1}{s} + \frac{1}{s'} = \frac{1}{f}$$

To illustrate this approach, consider the specific case where $f_1 = +3$ cm and $f_2 = -5$ cm. Take an object initially 4 cm from the lens combination and locate the final image. The calculation is simple as $s = +4$ cm and f (in cm) is obtained from:

$$\frac{1}{f} = \frac{1}{3} - \frac{1}{5} = \frac{2}{15}$$

Thus we have $f = +7.5$ cm, with for s' given by

$$\frac{1}{s'} = \frac{1}{7.5} - \frac{1}{4}$$

or s' = - 8.57 cm.

Location of this image graphically is shown in Fig. 34-9a, b and c, where the image I_1 of the first lens is treated as the "object" for the second lens in order to find the final image.

We will symbolize the thin lens combination by a vertical solid line. First we imagine just lens 1 to be located on that line and locate the image due to 1. See Fig. 34-9a.

We now imagine lens 2 to be on the vertical dashed line and treat I_1 as the object (note the object distance would be negative) for it. See Fig. 34-9b. For the combination see Fig. 34-9c.

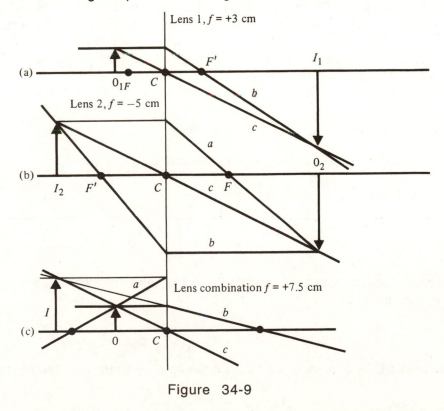

Figure 34-9

Example 12

A camera with focal length f must be able to image objects as close as 25 cm as well as those objects that are infinitely far away. How much "travel" of the lens is necessary if
(a) f = 35 mm; and (b) f = 50 mm?

Solution:

The thin lens equation is used to calculate the image distance (s') for the two extreme object distances, s = 25 cm and s = ∞. The difference between these two distances represents the "travel" required for the lens. The thin lens formula is:

$$\frac{1}{s} + \frac{1}{s'} = \frac{1}{f}$$

(a) Substituting f = 35 mm = 3.5 cm and s = 25 cm gives:

$$\frac{1}{s'} = \frac{1}{3.5 \text{ cm}} - \frac{1}{25 \text{ cm}}$$

Solving for s' yields s' = 4.07 cm.

When the object distance is infinite, then s' = f = 3.5 cm. For the image to strike the film in both cases, it must be possible to adjust the distance from lens to film by the amount Δs' where:

$$\Delta s' = 4.07 \text{ cm} - 3.5 \text{ cm} = 0.57 \text{ cm} = 5.7 \text{ mm}.$$

(b) Substituting f = 50 mm = 3.5 cm and s = 25 cm gives:

$$\frac{1}{s'} = \frac{1}{5.0 \text{ cm}} - \frac{1}{25 \text{ cm}}$$

Solving for s' gives s' = 6.25 cm so that Δs' has the value

$$\Delta s' = 6.25 \text{ cm} - 5.0 \text{ cm} = 1.25 \text{ cm} = 12.5 \text{ mm}.$$

Example 13

For a given object, distance, and light condition, what change in f-stop value is required to double the exposure time?

Solution:

A change of diameter D by a factor of =2 changes the area by a factor of 2 since the area can be written:

$$A = \pi R^2 = \frac{\pi D^2}{4}$$

Increasing the f-stop value *reduces* the aperture diameter and *decreases* the light reaching the film.   For typical f-stop values, moving up one "f-stop" decreases the diameter so that the area is halved requiring that the exposure time be doubled.

Example 14

A projector is required to image an 8.5 x 11 inch page onto a screen 20 ft away. The long side of the page (11 inches) is to be 5 feet high. What focal length is needed for the projector lens?

Solution:

The magnification needed is:

$$|M| = \left|\frac{s'}{s}\right| = \frac{5 \text{ ft}}{\left(\frac{11}{12} \text{ ft}\right)} = \frac{60}{11} = 5.455$$

Since the image distance is given as 20 ft, the object distance can be found from the magnification,

$$\left|\frac{20 \text{ ft}}{s}\right| = 5.455$$

or s = 3.67 ft. The focal length (f) is then found from the thin lens equation:

$$\frac{1}{s} + \frac{1}{s'} = \frac{1}{f}$$

where s and s' are known.

$$\frac{1}{f} = \frac{1}{3.67 \text{ ft}} + \frac{1}{20 \text{ ft}} = \frac{20 + 3.67}{(20)(3.67) \text{ ft}} = \frac{0.323}{\text{ft}} = \frac{1}{3.10 \text{ ft}}$$

Example 15

The "near point" of a person's eyes changes from 45 cm to 125 cm. What power lens (in diopters) is needed to correct the vision from the actual near point to the nominal 25 cm point in front of the eye?

Solution:

The "power" of a lens is the reciprocal of its focal length expressed in meters. The desired focal length is found from the thin lens equation:

$$\frac{1}{s} + \frac{1}{s'} = \frac{1}{f}$$

where s and s' are known to be 25 cm and - 45 cm respectively.

$$\frac{1}{f} = \frac{1}{25 \text{ cm}} + \frac{1}{-45 \text{ cm}} = \frac{25 - 45}{(25)(-45) \text{ cm}} = \frac{4}{2.25 \text{ m}} = 1.778 \text{ diopters}$$

The actual value of f was; f = 56.25 cm.

If the "near point" is 125 cm, then s = 25 cm and s' = - 125 cm.

$$\frac{1}{f} = \frac{1}{25 \text{ cm}} + \frac{1}{-125 \text{ cm}} = \frac{25 - 125}{(25)(-125) \text{ cm}} = \frac{4}{1.25 \text{ m}} = 3.20 \text{ diopters}$$

For two thin lenses (with focal lengths f_1 and f_2), in contact, the reciprocal of the focal length of the combination is the sum of the reciprocals of the two thin lens focal lengths--so the power adds. Thus if the person already had a lens with power 1.78 diopters and wanted a lens with power of 3.2 diopters, it would be possible to place a lens with power 1.42 diopters in contact with the first lens.

Example 16

A farsighted person has 45 diopters power when viewing distant objects but the image forms 2 mm in front of the retina. What power is needed for a corrective lens?

Solution:

Using the thin lens equation

$$\frac{1}{s} + \frac{1}{s'} = \frac{1}{f}$$

we set s = ∞ and s' = d - 2 x10⁻³ m where d is the effective lens-to-retina distance, and 1/f = 45 diopters. Thus

$$s' = d - 2 \times 10^{-3} \text{ m} = \frac{1 \text{ m}}{45} = 0.0222 \text{ m}$$

yielding d = 24.2 x10⁻² m. The power needed is d⁻¹ since the object distance is infinity

$$P = \frac{1}{d} = \frac{1}{.0242 \text{ m}} = 41.3 \text{ diopters}$$

Lenses are ground to the nearest 1/4 diopter, so a lens of (41.25-45) diopters = - 3.75 diopters is needed to correct this vision defect. The corrective lens is a diverging lens.

Example 17

Suppose a magnifying glass is held so that the image is produced a distance D from the eye. Calculate the angular magnification (M) under these conditions.

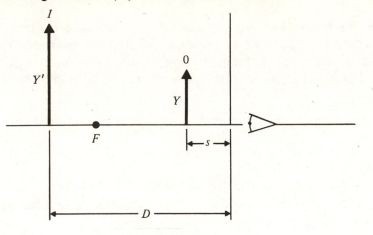

Figure 34-10

Solution:

A sketch of this magnifier is given in Fig. 34-10. The object (O) is located inside F so the image (I) is erect, enlarged, and virtual. The distance D is equal to - s'. If we view the object directly, at 25 cm away from the lens of the eye, then the angle subtended at the eye, u, is equal to

$$u = Y/25$$

where Y, Y', f and D are in cm. If we place our eye very close to the lens, the angle subtended at the eye by the image, u', is approximately equal to:

$$u' = Y'/D$$

Thus M = u'/u is given by

$$M = \frac{Y'}{Y}\left(\frac{25}{D}\right)$$

The ratio Y'/Y is equal to -s'/s, so to calculate it write:

$$\frac{1}{s} = \frac{1}{f} - \frac{1}{s'} = \frac{1}{f} + \frac{1}{D} \qquad \text{since } - s' = D.$$

Then

$$\frac{-s'}{s} = - s'\left(\frac{1}{f} + \frac{1}{D}\right) = D\left(\frac{1}{f} + \frac{1}{D}\right)$$

237

This gives for $M = (- s'/s)(25/D)$ the value:

$$M = D\left(\frac{1}{f} + \frac{1}{D}\right)\left(\frac{25 \text{ cm}}{D}\right) = \frac{25 \text{ cm}}{f} + \frac{25 \text{ cm}}{D}$$

The expression in the text was derived by assuming the image is located at infinity. The normal eye can focus on an image located as close as 25 cm, but no closer. Thus the angular magnification (subject to being able to focus the eye) is maximum for $D = 25$ cm and equal to $1 + 25/f$ in that case. If D is set equal to infinity, the above expression gives the same result as in the text.

Example 18

A microscope has an objective lens with focal length $f_o = 1.6$ cm and an eyepiece lens with focal length $f_e = 2.5$ cm. What is the magnification of the microscope if an object is placed 0.10 cm from the focal point of the objective lens?

Solution:

The magnification of a compound microscope is approximately equal to:

$$M_{total} = \frac{(25 \text{ cm})s'_1}{f_o f_e}$$

where the focal lengths and s_1' are expressed in centimeters. The image distance s_1' is calculated from the thin lens equation given $f_o = 1.6$ cm and $s = 1.6$ cm + 0.10 cm = 1.7 cm:

$$\frac{1}{s_1'} = \frac{1}{f} - \frac{1}{s_1} = \frac{1}{1.6 \text{ cm}} - \frac{1}{1.7 \text{ cm}} = \frac{0.1}{(1.6)(1.7) \text{ cm}}$$

Therefore $s_1' = 27.2$ cm. This yields for the magnification,

$$M = \frac{(25 \text{ cm})(27.2 \text{ cm})}{(1.6 \text{ cm})(2.5 \text{ cm})} = 170.$$

Example 19

In a simple telescope, the image is inverted with respect to the object. Suppose a third lens is used to invert the image of the objective lens without enlarging or reducing the image. How much longer must this telescope be?

Solution:

Use the approach for an optical system. The first image becomes the object for the next lens, etc. The third lens is called the erecting lens. Let f be the focal length of this lens. To have no magnification from this lens, s' and s must have the same magnitude. To invert the image, they must have the same sign, so s = s'. Thus

$$\frac{1}{s} + \frac{1}{s'} = \frac{1}{f} \quad \text{since } s = s' \quad \text{gives } s = 2f = s'.$$

If this erecting lens is added between the objective lens and the eyepiece, the image of the objective lens must fall a distance s from the third lens and the eyepiece must view the (now erect) image at a distance s' from this lens. Therefore the telescope must be s + s' = 4f longer than it was originally.

QUIZ

1. A spherical mirror with radius of curvature of 0.4 m forms an image of a 2 cm high object located 1.2 m from the vertex of the mirror. (a) Calculate the position and height of the image if the mirror is concave.(b) Calculate the position and height of the image if the mirror is convex.

Answer: (a) s' = 0.24 m, y' = - 0.40 cm; (b) s' = - 0.171 m, y' = 0.286 cm

2. A glass rod (n_g = 1.55) is submerged in water (n_w = 1.33). The end of the rod is spherical (convex) with radius of curvature R = + 0.20 m.
    (a) How far from the vertex should an object be placed (in water) to produce an image at infinity?
    (b) Where will the image of an infinitely distant object in water be located?

Answer: (a) s = 1.21 m; (b) s' = 1.41 m (inside the glass)

3. An object in water (n_w = 4/3) is located 1 m to the left of a concave spherical refracting surface with index of refraction n_g = 1.50. The magnitude of the radius of curvature is 2 m. Using the small angle approximation, locate the position of the image.

Answer: The image is 1.059 m to the left of the refracting surface.

4. A hemispherical piece of glass (with refractive index of 1.50) is placed over an object on a flat surface. The small object is at the center of the circular bottom surface of radius R = 20 cm. Use the small angle approximation to locate the image of the object.

Answer: The image is located at the same position as the object.

5. An object of height 3 cm is located + 20 cm from a thin lens with index of refraction 1.55. Calculate the position of the image and its height if:
(a) R_1 = +10 cm and R_2 = -20 cm; and (b) R_1 = -10 cm and R_2 = + 20 cm

Answer: (a) s' = + 30.8 cm and y' = - 4.62 cm; (b) s' = - 7.55 cm and y' = + 1.13 cm

6. Two thin lenses with f_1 = + 15 cm and f_2 = + 25 cm are located 5 cm apart. Calculate the location of the final image and the lateral magnification.

Answer: The image is 14.13 cm from the second lens (19.13 cm from the first lens) and the magnification is + 0.65.

7. A real object is placed 8 cm in front of a converging lens with f_1 = + 6 cm. A second diverging lens with f_2 = - 4 cm is placed 4 cm behind the first lens. Find the location of final image with respect to the second thin lens.

Answer: The image is located 5 cm in front of lens 2 (1 cm in front of lens 1).

8. If the film in a 35 mm camera is 24 mm in height, how tall an object will it take to completely fill the film if the object is 80 cm from the lens?

Answer: 52.4 cm (or 20.6 inches).

9. What must be the focal length of the eye lens to form an image 2 cm behind the lens of an object: (a) 25 cm in front of the eye? (b) of an infinitely distant object?

Answer: (a) f = 1.852 cm; (b) f = 2 cm.

10. If a microscope has f_o = 1.6 cm and f_e = 2.5 cm, what is the distance between the two lens for an object viewed 0.15 cm in front of the focal point of the objective lens and what is its magnification?

Answer: d = 20.94 cm and M = 117.

11. An object of height 1 cm is viewed by the unaided eye, 25 cm away from the eye. A magnifying glass of focal length f = + 10 cm is used to view the object. If the final image, after magnification, is also 25 cm away from the eye, calculate the angular magnification in the small angle approximation.

Answer: 3.5

35

INTERFERENCE

OBJECTIVES

In this chapter waves of the same frequency, possessing a constant phase relationship, will produce an interference pattern when they overlap in space. This pattern has alternate light and dark regions called fringes. Your objectives are to:

Define the concept of coherence as a constant phase relationship.

Obtain and apply the relationship between phase difference and optical path length.

Incorporate the effect of phase shifts introduced by certain reflections.

Calculate the main features of various interference patterns.

Huygens' principle, implying that each wavefront is a source of secondary wavelets, is central to the arguments of this chapter. The critical derivation for the intensity of the pattern observed for two slit interference is obtained by adding phasors, encountered in previous chapters, and then squaring the resultant vector to obtain the intensity.

REVIEW

Electromagnetic waves of the same frequency can interfere with each other if they overlap spatially. This interference can be of either the constructive or destructive type. In constructive interference, the total intensity is larger than the sum of the individual intensities from the various sources, whereas in destructive interference the resultant intensity is less than the sum of the individual intensities from the various sources. Although the interference pattern always exists in principle, in practice it may not be observable or detectable. In order for this pattern to be detectable (by ordinary means such as the human eye or photographic film) the phase difference between the two (or more) sources contributing to the pattern must not change appreciably in the time interval over which the pattern is observed. In this case, the sources are said to be *coherent*. Thus, if conventional methods of detection are used to observe the interference between two sources, the time over which the sources are coherent must be larger than the time constant that characterizes the detector (about 0.1 s for the eye and 10^{-3} s for fast film).

The topics covered by this review include: Young's experiment; thin-film interference; Newton's Rings; the Michelson interferometer; and the photon. A summary paragraph of the treatment of interference is also included.

YOUNG'S EXPERIMENT

The classic example of an experiment that demonstrates the interference of light waves from two coherent sources is Young's experiment. If a single, monochromatic light source S is used to illuminate two slits, the slits S_1 and S_2 can be regarded as sources of secondary wavelets (in the Huygens sense) and act as two *coherent, monochromatic,* sources of light. The coherence of S_1 and S_2 arises from the fact that they are derived from the same source S so that a change in the phase constant of source S in time is passed along equally to S_1 and S_2 making their difference in phase constants equal to zero for all times.

The alternation of light and dark regions observed, say in Young's experiment, is called an *interference pattern.* If we blocked off one of the sources, say S_2, we would expect to see a region of geometric brightness--with uniform illumination--and a region of geometric shadow outside the bright region. The light pattern obtained in the two slit Young's experiment consists of light and dark interference fringes both in the region of geometric brightness and in the geometric shadow.

THIN FILM INTERFERENCE

The transverse electromagnetic waves that we call light are solutions of the wave equation and hence will suffer the same phase changes upon reflection as waves or pulses on a string. The wave velocity in optics is incorporated into the refractive index. Maxwell's equations, applied to the problem of reflection at an interface, give the desired solution but we will just state the result without proof. If the electromagnetic wave strikes an interface with index of refraction higher than that of the medium in which it was traveling, then the reflected wave suffers a phase change of π radians with respect to the incident wave. If the wave strikes an interface with lower index of refraction than the medium in which it was traveling, no phase change upon reflection occurs. This very important rule is illustrated in Example 2 and then used in Example 3 to obtain the interference pattern known as Newton's Rings. These phase changes that can occur upon reflection play an important role in the interference effects that can be observed with thin films and special mirrors (Lloyd's mirrors).

NEWTON'S RINGS

The central region in a Young's experiment is *bright*, because the path differences and hence the overall phase difference is zero. The interference pattern known as "Newton's Rings", however, has a *central dark region* just where the path difference, $r_1 - r_2$, is zero. Similarly in the interference phenomena observed with Lloyd's mirror the fringe observed where the actual path length is zero is dark. (One of the two beams, however, suffers a glancing reflection from a mirror surface.) To explain these phenomena, it is necessary to use the idea given in the previous paragraph, that a phase shift of π radians occurs for some types of reflections but not for others. This is illustrated in Example 3.

A SUMMARY OF THE TREATMENT OF INTERFERENCE

To summarize the essential ideas for our treatment of interference:

(1) The problem of making two *spatially separated* sources of electromagnetic waves coherent over long time intervals is solved by using one source and somehow deriving two (or more) beams from it that in turn are used for the sources to set up the interference pattern.

(2) For two "sources" of the above type with equal amplitudes, the light intensity at all points in the pattern depends only on the relative phase of these two sources. Included in this relative phase are differences in optical path length (real path length multiplied by index of refraction) and phase changes suffered upon certain reflections.

(3) When this total phase difference is π, 3π, 5π, etc., the waves interfere destructively and the intensity is zero. When this phase difference is 0, 2π, 4π, etc., the waves interfere constructively and the intensity is four times larger (for two sources) than the intensity due to one source alone. The average intensity over the entire pattern is just twice that expected from a single source.

(4) The critical equation for understanding all of these results is

$$I = I_0\cos^2(\delta/2)$$

where δ is the total difference in relative phases. This equation was derived in the text by finding the vector resulting from the addition of two vectors with the same amplitude but shifted in relative phase by the amount δ (phasors). *This is a very important result.* To calculate δ for our two sources at a given point we need to know the *difference in optical path lengths* of the two beams in getting to the point *plus the accumulation of all the π phase shifts upon reflection*. These notions are discussed in Example 2.

It is interesting to note that the constant I_0 represents an intensity that is *four* times larger than the intensity of a single slit source. Thus at the center of the pattern, the intensity is I_0. If the two sources S_1 and S_2 contributed independently to the illumination of that point *without interference*, we would obtain $I(y=0) = 2 I_1 = I_0/2$. The center spot thus has double the intensity we would expect for two sources *without* interference. This however is balanced out by the existence of regions with no intensity (dark fringes), so that on average, the average intensity on the screen works out to be $2 I_1$. We neither gain nor lose intensity by allowing the sources to interfere with each other.

THE MICHELSON INTERFEROMETER

A.A. Michelson designed an experimental apparatus capable of precise length measurements or length change measurements using the interference of light. Such a device is now called an interferometer. Michelson also used interferometer techniques in collaboration with Morley (the Michelson-Morley experiment) to perform one of the most famous "null" experiments in physics. In this experiment, with one of the beams parallel to the direction of the earth's

243

orbital motion about the sun, the interferometer was rotated through 90°. No shift in the interference fringe pattern was observed hence the experiment gave a "null" result. However, according to the accepted theories of the day regarding propagation of electromagnetic waves, there *should* have been a shift.

THE PHOTON

Einstein first postulated the light was composed of discrete packets of energy where the energy (E) in the packet was related to the frequency (ν) of the light by:

$$E = h\nu$$

where $h = 6.63 \times 10^{-34}$ J·s was a universal constant, called Planck's constant. With this postulate, Einstein was able to explain a puzzling experimental result known as the "photoelectric effect". This postulate has been amply confirmed by experiment and is one of the "cornerstones" of Modern Physics. The packets of energy (or quanta) are now called *photons* and light is endowed with "particle-like" properties.

The photon is unlike any other particle we have studied in that it is massless ($m_{photon} = 0$) and always travels with the speed of light! Its properties are correctly described by Einstein's theory of Special Relativity (discussed in Chapter 37). When that theory is applied to massless particles, it is found that the relationship between energy and linear momentum is:

$$E = pc$$

where c is the speed of light (of the photon). Since the wavelength, frequency and wave speed are related by $\lambda = c/\nu$, the linear momentum of the photon is given by:

$$p = \frac{h}{\lambda}$$

where (again) h is Planck's constant. Thus beams of light carry not only energy but linear momentum. The intensity of a monochromatic light beam is proportional to the number of photons in that beam. A collision between a photon and an electron is treated in Example 8.

HINTS AND PROBLEM SOLVING STRATEGIES

Make a sketch as before. Identify the two (or more) sources, S_1, S_2, etc. Find the phase or path difference between these sources at the point of observation. Remember that certain reflections can produce phase changes. After calculating the net phase change, due to optical path length differences and reflection phase changes, between the sources, use the rules given for interference patterns to find the maxima and minima.

QUESTIONS AND ANSWERS

Question. A thin film of soapy solution on a wire is viewed by reflected light. Why does the thinnest section of the film appear black?

Answer. There is an interference between a wave reflected from the front and the back of the soap film, which are out of phase by π radians, completely canceling each other.

Question. Newton's rings are formed when a glass lens rests on a flat piece of glass in air, as shown in Fig. 35-3. Suppose the medium between the lens and the glass slab has a higher index of refraction than the glass. Is the center of the pattern still dark?

Answer. Yes. There is still a *relative* phase change of π radians between the waves reflected by the lens and the flat, although in the opposite direction.

EXAMPLES AND SOLUTIONS

Example 1

Calculate the separation between the interference fringes obtained in Young's experiment when $\lambda = 550$ nm, $R = 3$ m, and $d = 0.22$ mm. Refer to Fig. 35-1.

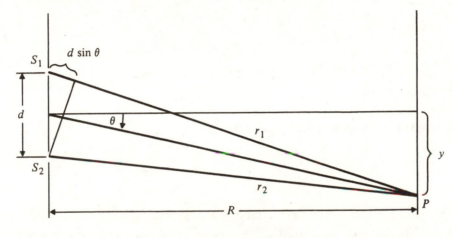

Figure 35-1

Solution:

From the sketch given in Fig. 35-1, the path difference ΔD, from the two coherent sources to point P, in the small angle approximation, is given by (refer to the derivation of this result in the main text):

$$\Delta D = d \sin \theta$$

When this path difference is equal to an integer multiplied by the wavelength, we get *constructive interference.*

$$\Delta D = n\lambda \qquad n = 0, 1, 2, \dots \quad \text{(bright fringe)}$$

Conversely when the path difference is equal to a half-integral number of wavelengths we get *destructive interference*

$$\Delta D = \left(\frac{2n + 1}{2}\right)\lambda \qquad n = 0,1,2,... \quad \text{(dark fringe)}$$

Since the distance to the screen (R) is large compared to y, the sine can be approximated by

$$\sin \theta \cong \tan \theta = (y/R)$$

Thus bright fringes appear for values of y (labeled y_n)

$$y_n = \left(\frac{n\lambda}{D}\right)R$$

Dark fringes appear for values of y_n satisfying

$$y_n = \left(\frac{2n + 1}{2}\right)\frac{\lambda}{D}R$$

Numerically we have

$$\frac{\lambda}{D}R = \frac{(5.50 \times 10^{-7} \text{ m})(3 \text{ m})}{(0.22 \times 10^{-3} \text{ m})} = 7.5 \times 10^{-3} \text{ m} = 7.5 \text{ mm}.$$

The spacing between like fringes is then 7.5 mm.

Example 2

Let vectors $\vec{E}_1$ and $\vec{E}_2$ represent the electric fields at the point of observation P from slits 1 and 2 respectively. (a) If the light from slit 2 travels a longer path distance than that from slit 1 by the amount $\Delta r = d \sin \theta$, show that the phase difference between $\vec{E}_1$ and $\vec{E}_2$ is

$$\Delta \Phi = \frac{2\pi}{\lambda}(d \sin \theta).$$

(b) Let $\vec{E}_1$ and $\vec{E}_2$ be vectors of the same length, E, but with $\vec{E}_2$ making an angle ϕ with $\vec{E}_1$.
   Calculate the magnitude of the resultant vector $\vec{E}_R = \vec{E}_1 + \vec{E}_2$.
(c) Assuming the intensity is proportional to E_R^2, find the intensity on the screen.

Solution:

(a) The phase difference $\Delta \Phi$ is related to the path difference Δr by

$$\frac{\Delta \Phi}{2\pi} = \frac{\Delta r}{\lambda}$$

Thus if $\Delta r = d \sin \theta$ for two extremely narrow slits separated by distance d, then

$$\Delta \Phi = \frac{2\pi}{\lambda} (d \sin \theta).$$

(b) The magnitude of the resultant vector is the sum of the squares of the x and y components,

$$E_R^2 = (E + E \cos \phi)^2 + (E \sin \phi)^2$$

$$E_R^2 = 2E^2 + 2E^2 \cos \phi = 2E^2 (1 + \cos \phi) = 4E^2 \left(\cos^2 \frac{\phi}{2} \right)$$

The magnitude of the resultant is found by taking the positive square-root:

$$E_R = 2E \left(\cos \frac{\phi}{2} \right)$$

(c) Since the intensity I is proportional to the square of the resultant electric field vector we have:

$$I = I_0 \left[\cos^2 \left(\frac{2\pi \, d \sin \theta}{2\lambda} \right) \right] = I_0 \left[\cos^2 \left(\frac{\pi \, d \sin \theta}{\lambda} \right) \right]$$

where $I_0 = 4E^2$. The pattern is bright when the argument of the cosine is 0, π, 2π, etc. and the intensity of the nth bright region is equal to that at the center.

Example 3

Compare the "phase shifts" and path differences of rays A, B and C in Fig. 35-2 for <u>normal incidence</u> of the incident ray I. Use the phase of I at point P as the reference phase.

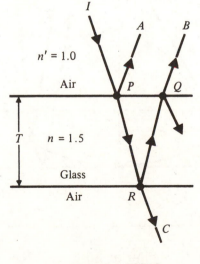

Figure 35-2

247

Solution:

Here the incident ray, I, as shown in Fig. 35-2, is drawn approaching the surface at a non-zero angle of incidence for convenience in following the various rays. In the end we will let the angle of incidence be zero so points P and Q will coincide. Ray A is the part of the incident beam reflected at P. Since it is coming from air and reflected from glass, at point P. The phase shifts are as follows.

Ray A: As this ray is reflected from a material with index higher than that of the original material, it suffers a phase change of π radians.

Ray B: This ray is reflected from a material with lower index than the original material so there is no phase shift. Also there is no phase shift upon transmission (ever). This ray has traveled an additional distance (for normal incidence) of 2T *in glass* compared to the incident ray. The additional *optical path* is then 2nT where n is the index of refraction of the glass. With respect to the phase of I at point P, the phase change of ray B is:

$$\Delta \Phi = \frac{2\pi}{\lambda}(2nT)$$

Ray C: This ray is transmitted at both the top and bottom surfaces and hence suffers no phase change. The optical path length through the glass is nT. If this wave was compared to the part of the incident wave that *did not* pass through the glass, the optical path difference of these two waves would be

$$\Delta D = nT - T = (n - 1)T$$

Example 4

Develop a formula for the radii of both the dark and light rings in the Newton's Rings pattern.

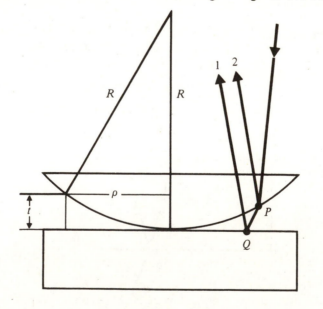

Figure 35-3

Solution:

Referring to Fig. 35-3, the distance P, from the center out to the point where the thickness of the air space is t, is found from

$$R^2 = (R-t)^2 + \rho^2 = R^2 - 2Rt + t^2 + \rho^2$$

$$(2R - t)t = \rho^2$$

If R >> t then the thickness t is related to ρ by

$$t = \frac{1}{2R}\rho^2$$

Consider now two possible paths for a light ray. Rays along these two paths are coherent because they originate at a single source. These paths are labeled 1 and 2 in the above figure. For path 2, the ray is reflected at point P *in glass* and hence suffers no phase shift. For ray 1, the reflection occurs at Q *in air* so there is a phase change of π there. In addition, ray 1 has a longer optical path, by an amount 2t (if viewed normally) where t is the thickness of the air gap at that point. Usually we would say that when the path difference was equal to an integral number of wavelengths we would get constructive interference (and hence bright fringes) but the relative phase change of π now means that when 2t (the path difference between rays 1 and 2) equals an integral number of wavelengths, we will get destructive interference and hence dark fringes.

$$(\text{dark}) \quad 2t = n\lambda = \frac{2}{2R}\rho_n^2 \quad n = 0, 1, 2, ...$$

Thus the radius of the nth dark ring, ρ_n, is equal to

$$(\text{dark}) \quad \rho_n = (nR\lambda)^{1/2} \quad n = 0, 1, 2, ...$$

where R is the radius of curvature of the lens.

To get constructive interference, we must make the path difference a half-integral number of wavelengths, or

$$(\text{light}) \quad 2t = \left(\frac{2n - 1}{2}\right)\lambda.$$

$$(\text{light}) \quad \rho_n = \sqrt{\left(\frac{2n - 1}{2}\right)R\lambda} \quad n = 1, 2, 3, ...$$

Note the center is dark since the air thickness is negligible there but the two waves have a relative phase difference of π.

Example 5

A thin layer of water ($n_W = 4/3$) when viewed normally produces a "non-glare" optical coating on glass ($n_G = 1.5$) for light of wavelength 600 nm. What is the minimum thickness of the layer of water?

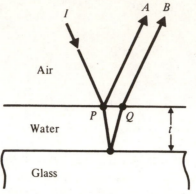

Figure 35-4

Solution:

Referring to Fig. 35-4, ray A suffers a phase change of π upon reflection at P (since it is trying to get into a medium with higher index of refraction). Ray B also suffers a phase change of π when reflected at point R since glass has a higher refractive index than water. To make this "non-glare" coating, we want rays A and B to interfere destructively. For this to occur, the optical path difference must be equal to a half integral number of wavelengths. The path difference, for normal viewing, is 2t but the *optical path difference* equals $2tn_W$. For minimum thickness, this difference should be equal to one half wavelength:

$$2tn_W = \lambda/2$$

$$t = \lambda/4n_W$$

Numerically we have

$$t = \frac{3}{4}\left(\frac{600 \times 10^{-9} \text{ m}}{4}\right) = 112.5 \text{ nm}$$

Example 6

A soap film with refractive index n = 1.35 is viewed normally. Calculate the minimum thickness of the film (d) that will give constructive interference for yellow light of wavelength $\lambda = 575$ nm.

Solution:

For constructive interference, the minimum optical path difference must be equal to one-half wavelength ($\lambda/2$) since a phase change of π radians occurs upon reflection from the top surface. The optical path difference (Δp) is the product of the geometric path difference (Δr) and the index of refraction.

$$\Delta p = n(\Delta r)$$

The geometric path difference is $\Delta r = 2d$ where d is the film thickness, therefore:

$$\Delta p_{min} = n \Delta r_{min} = n(2d_{min}) = \frac{\lambda}{2}$$

Solving for the minimum thickness (d) by substitution of the numerical values given, we have:

$$d_{min} = \frac{1}{4}\frac{\lambda}{n} = \frac{1}{4}\left(\frac{5.75 \times 10^{-7} \text{ m}}{1.35}\right) = 1.065 \times 10^{-7} \text{ m}$$

Example 7

Using the "classical addition law of relative velocities, calculate the fringe shift expected for a Michelson Interferometer when rotated from an orientation where light travels parallel to the earth's velocity to one where it travels perpendicular to the earth's velocity. Assume the wavelength of light used is $\lambda = 590$ nm and the length L of the path is 11 m.

Solution:

In Michelson's time it was thought that a medium called the ether (aether) filled space and made possible the propagation of electromagnetic waves just as a medium was required for the propagation of sound waves. The speed of light, c, was considered to be the speed of light relative to the ether. The earth moves about the sun with speed $v_e \cong 3 \times 10^4$ m/s (again with respect to the ether). The speed of light has the value $c = 3 \times 10^8$ m/s so that the ratio $v_e/c = 10^{-4}$ and is small compared to unity.

The two beams in the interferometer each have phase shifts upon reflection that total 2π radians. A flat glass plate acts as a "compensator" to make the optical path through glass the same for both beams. When we align the interferometer so that the beams travel parallel to the earth's motion, the time, T_2, for the beam to traverse the distance L_2 and return to the beam splitter is

$$T_2 = \frac{L_2}{c - v_e} + \frac{L_2}{c + v_e} = \frac{2L_2 c}{c^2 - v_e^2}$$

where the relative velocities are taken to be $c - v_e$ and $c + v_e$ depending on the direction of the beam with respect to the earth's velocity.

When the direction of propagation of light is perpendicular to the earth's velocity, the time T_1 to traverse the path L_1 is then

$$T_1 = \; = \frac{2L_1}{\sqrt{c^2 - v_e^2}}$$

The interferometer used by Michelson and Morley had equal length arms so $L_1 = L_2 = L$. The predicted time difference between these two paths T is

$$T = T_2 - T_1 = \frac{2Lc}{c^2 - v_e^2} - \frac{2L}{\sqrt{c^2 - v_e^2}}$$

$$T = \frac{2Lc}{c^2 - v_e^2}\left(1 - \sqrt{1 - \frac{v_e^2}{c^2}}\right)$$

Since $(v_e/c)^2 = 10^{-8}$ and is very small, we can approximate this time difference by

$$T \cong \frac{2Lcv_e^2}{2c^4} = \frac{Lv_e^2}{c^3}$$

Therefore by rotating the interferometer through 90°, a time difference of T and a path difference of $c(2T)$ is created. The fringe shift, ΔN, that results from this path difference is then

$$\Delta N = \frac{c(2T)}{\lambda} = \frac{2c}{\lambda}\left(\frac{Lv_e^2}{c^3}\right) = \left(\frac{2Lv_e^2}{\lambda c^2}\right)$$

Numerically, using $\lambda = 590$ nm and $L = 11$ m, the predicted shift $\Delta N = .37$ fringe. No fringe shift was observed although the spectrometer was designed to observe a fringe shift as small as .04 fringe. This "null" experimental result was exceedingly important to the theory of relativity gaining scientific acceptance.

Example 8

A photon (wavelength λ) traveling in the + x direction collides with an electron at rest of mass m. The photon rebounds in the - x direction after a completely elastic collision with the electron. Calculate the change in photon wavelength $\Delta\lambda$, assuming that this change is small compared to the original wavelength.

Solution:

For a completely elastic collision, both linear momentum and kinetic energy are conserved. The energy (E) of the photon is all kinetic energy. Therefore kinetic energy conservation yields:

$$h\nu = h\nu_f + \frac{1}{2} mv^2$$

Since kinetic energy is transferred to the electron as a result of the collision, the photon energy (and frequency--hence "color") changes!
Conservation of linear momentum gives:

$$\frac{h}{\lambda} = -\frac{h}{\lambda_f} + mv$$

where $\lambda_f = c/\nu_f$ is the "new" wavelength, related to the original wavelength by $\lambda_f = \lambda + \Delta\lambda$. The remainder of the problem involves completing the algebraic solution for the approximate change in wavelength $\Delta\lambda$. Re-arranging the kinetic energy equation,

$$h\frac{c}{\lambda} = h\frac{c}{\lambda_f} + \frac{1}{2} mv^2$$

We can solve the above two equations for v and v^2;

$$\frac{h}{m}\left[\frac{1}{\lambda} + \frac{1}{\lambda_f}\right] = v$$

$$\frac{2hc}{m}\left[\frac{1}{\lambda} - \frac{1}{\lambda_f}\right] = v^2$$

Combining these results in:

$$\frac{2hc}{m}\left[\frac{1}{\lambda} - \frac{1}{\lambda_f}\right] = \left(\frac{h}{m}\left[\frac{1}{\lambda} + \frac{1}{\lambda_f}\right]\right)^2$$

Using $\lambda_f - \lambda = \Delta\lambda$, we obtain:

$$\Delta\lambda = \frac{h}{2mc}\left[\frac{(2\lambda + \Delta\lambda)^2}{\lambda(\lambda + \Delta\lambda)}\right]$$

This is still exact. To approximate, let $2\lambda + \Delta\lambda \cong 2\lambda$ and $\lambda + \Delta\lambda \cong \lambda$ on the right hand side. The final expression is:

$$\Delta\lambda \cong \frac{2h}{mc} .$$

QUIZ

1. In the geometry of Young's experiment, shown in Fig. 39-1, the two slits are separated by 0.1 mm, the distance to the screen is 0.6 m, and the wavelength of the source used to illuminate the slits is $\lambda = 500$ nm. Find the position of the second minimum away from the central maximum.

Answer: $y = 4.5 \times 10^{-3}$ m (symmetric about the central maximum).

2. A lens with an index of refraction n = 1.70 is coated with a film of index n_f = 1.40. (a) Calculate the minimum, non-zero, thickness of the film if a wavelength of 550 nm is to interfere destructively with itself (on reflection) when incident from air on the film at an angle of 0°. (Part of the light is reflected at the air-film interface and part is reflected at the film-glass interface). (b) Calculate the minimum thickness of this film if two wavelengths, λ_1 = 550 nm and λ_2 = 450 nm are to each interfere destructively with itself. This is a non-reflective coating for two wavelengths.

Answer: (a) 9.82×10^{-8} m   (b) 8.84×10^{-7} m.

3. Using a Michelson interferometer it is possible to measure very small length changes with high accuracy. In an interferometer with equal length arms L = 1 m, a glass slide of refractive index n = 1.5 and thickness 1 mm is inserted into one arm. Take c = 3 x 10^8 m/s and the wavelength of light used to be 500 nm. (a) Calculate the change in optical path introduced for light transmitted through the glass slide twice (going and coming). (b) Calculate the fringe shift produced by introduction of that slide.

Answer: (a) $\Delta p = 3 \times 10^{-3}$ m; (b) $\Delta N = 6000$.

4. A laser with continuous power of 2 Watts and wavelength $\lambda = 550$ nm produces photons concentrated in a narrow beam in the + x direction. The beam strikes a blackened disk that absorbs all the incident photons. (a) What force does the photon beam exert on the disk? (b) Is it necessary to know the wavelength?

Answer: (a) $F = 6.67 \times 10^{-9}$ N in + x direction; (b) No.

5. An apparatus set up to produce Newton's Rings produces the first bright ring with radius ρ_1 = 1 mm when a source with wavelength λ_1 = 550 nm is used. What wavelength would be required to produce the first dark ring (outside the center) at this radius (of 1 mm)?

Answer: $\lambda_2 = 275$ nm

36
DIFFRACTION

OBJECTIVES

In our treatment of interference in the last chapter, it was seen that two or more coherent sources could add constructively or destructively, producing interference patterns characteristic of the system's geometry. The classic example is Young's experiment where two slits are illuminated by a single source. In this chapter, we see that there are observable interference effects even if there is only one slit.

Your objectives are to:

Compute the intensity of single and multiple slit interference patterns.

Estimate the resultant intensity using Fresnel zones.

Apply the concept of diffraction (of x-rays) to crystals.

Calculate the limits that diffraction places on the ultimate resolution of optical instruments.

REVIEW

In our analysis of Young's experiment, we took the slits to be infinitely narrow so that they were a line of point sources. We consider now the effect of finite slit width. Two approaches to calculating diffraction effects evolved historically leading to what is now called Fraunhofer diffraction or Fresnel diffraction named after their proponents.

FRAUNHOFER AND FRESNEL DIFFRACTION

For *Fraunhofer diffraction*, the monochromatic light source S is removed to optical infinity by placing it at the focal point, F_1, of a converging lens. The rays that illuminate the aperture (or single slit) are then parallel rays. By placing a second lens such that the center of the slit is at its focal point F_2, the rays reaching the observation screen will also be parallel rays. Thus in Fraunhofer diffraction, both the waves incident on the aperture or obstacle and the waves reaching the point of observation (the screen in this case) are plane waves. This simplifies the analysis.

For *Fresnel diffraction*, the obstacle is close enough to the source that the wavefronts are spherical. Furthermore, the detected wavefronts are again spherical since the detector is also close to the aperture or obstacle. We will concentrate on Fraunhofer diffraction as the analysis is somewhat simpler although we will use the concept of Fresnel zones in many of our analyses.

SINGLE SLIT DIFFRACTION

The coherence problem encountered with interference does not occur with diffraction effects which result from interference *between the various parts of the same slit*. If parallel light from a single slit with width comparable to the wavelength of light is collected by a converging lens and focused on a screen, the central region is bright but dark fringes appear with regularity as you move in either direction away from the central maximum. The angle at which these dark fringes appear can be predicted exactly by the following argument:

(1) The wavefront emerging from the single slit when viewed at an angle θ with respect to the normal can be thought of as being divided into N equal zones or regions.

(2) Each region can be characterized by an electric field vector of the same magnitude as that of all the other zones but differing in phase by an amount $\Delta\phi$ (which depends on the point of observation) with the electric field vectors representing the adjacent zones (i.e. each of these electric field vectors is a phasor.)

(3) The resultant electric field vector at a given point of observation is obtained by summing these N phasors.

(4) The intensity at the point of observation is proportional to the square of the resultant electric field.

To do the calculation exactly, the number of zones, N, is made infinitely large so that the sum of the phasors becomes the chord of an arc of a circle. The resultant electric field vector is

$$E_T = S \frac{\sin(\delta/2)}{(\delta/2)}$$

where S is the arc length and represents the sum of all the phasors when they have no phase difference between individual phasors. The intensity when viewed normally ($\sin \theta = 0$) is I_0 and is proportional to S^2. Thus the intensity at any viewing angle, θ, $I(\theta)$ is given by

$$I(\theta) = I_0 \frac{\sin^2(\delta/2)}{(\delta/2)^2}$$

where δ is total the phase difference between the top and bottom of the slit. If the optical path difference is Δp, then this phase difference is

$$\delta = \frac{2\pi}{\lambda}(\Delta p)$$

The path difference between top and bottom of the slit is *a sin* θ where "a" is the slit width and θ is the viewing angle. From the above expression for the intensity, it is seen that the sine function has zeroes when the argument is an integral multiple of π. The condition for dark fringes (zero intensity) is then:

$$(\delta/2) = n\pi \quad \text{where } n = 1, 2, 3, \text{ etc.}$$

Substituting the previous value for $\delta = 2\pi/\lambda(a \sin \theta)$, one has for the various values of φ where diffraction minima occur, θ_n, the following equation:

$$\frac{1}{2}\left(\frac{2\pi}{\lambda}\right) a \sin \theta_n = n\pi$$

or in terms of the angle

$$\sin \theta_n = n(\lambda/a)$$

If y_n is the distance from the center of the pattern on the screen to the nth minimum and F is the focal length of the lens used, then in the small angle approximation is

$$\sin \theta_n = \frac{y_n}{F} = n\left(\frac{\lambda}{a}\right)$$

From the above expressions, it is easy to see that the angular width of the central bright spot is double that of any of the other maxima and that the positions of the maxima are only approximately half-way between the dark fringes. The intensities of the maxima fall off very rapidly. See Example 4.

THE DIFFRACTION GRATING

To understand the results from two or more slits in complete generality, one must use ideas both from interference and diffraction. The diffraction grating is made by ruling many identical thin slits into a glass or plastic plate. Before the advent of the laser, the diffraction grating was the chief means of obtaining monochromatic light. The "reinforcement" condition is

$$d \sin \theta = N \lambda \quad \text{where } N = 1, 2, 3, \text{ etc.,}$$

with d the spacing between slits. In this case, the integer N is called the "order" of the pattern. Frequently two or more orders will overlap, reducing the usefulness of the grating.

X-RAY DIFFRACTION

When constructing a diffraction grating where $(d \sin \theta) = n\lambda$, one needs the grating spacing (d) to be comparable to the wavelength in order to obtain large values of the angle θ. This was technically difficult in earlier times, but possible for optical wavelengths of order 500 nm but for higher energy radiation, such as x-rays, the shorter wavelengths of order 1 nm imposed a much more stringent condition on the spacing d.

257

Fortunately, the spacing between planes of atoms in a crystal lattice provides values of d, where d is now the spacing between planes of atoms, that *are* comparable with x-ray wavelengths. X-rays scattered from such parallel planes, incident at an angle θ with respect to the plane surface of (here θ is not measured with respect to the normal to the plane) constructively interfere when

$$2d \sin \theta = n\lambda \quad \text{where } n = 1, 2, 3, \ldots$$

This relationship is known as the "Bragg condition".

In practice, x-ray sources have a nearly continuous distribution of wavelengths over a broad spectrum of values and real crystals have many different sets of planes that, for a given orientation of the beam with respect to the crystal, satisfy the Bragg condition *in some order*. The resulting x-ray diffraction pattern can be very complicated but each type of crystal structure *has its own distinctive pattern* making x-ray diffraction a standard method of determining crystal structure.

LIMITS ON RESOLVING POWER

Diffraction effects limit the resolving power of most optical instruments. Usually the diffraction effect results from the finite size of some circular aperture. For a circular geometry, the intensity distribution <u>does not</u> follow the function

$$\frac{\sin^2(\delta/2)}{(\delta/2)^2}$$

as it does for rectangular geometry but rather is described by a Bessel function. The nulls of Bessel functions are <u>not</u> evenly spaced. The first null is located such that the radius of the first dark ring, R, is given by

$$R = 1.22 \left(\frac{\lambda}{D}\right) F$$

where F is the focal length of the lens used and D is the diameter of the limiting aperture (usually the lens). An objective criterion for deciding when the angle between two sources is resolvable is called the "Rayleigh criterion". Using this criterion, the sources are just resolvable when the maximum of one diffraction pattern falls at the minimum of the other. This is illustrated in Example 7.

FRESNEL ZONES

An interesting reconstruction of the wavefront into zones known as Fresnel zones makes it possible to estimate light intensities when diffraction effects occur with very simple physical arguments. We present that picture below. In Figure 36-1, a circular aperture is shown, uniformly illuminated by a monochromatic source to the left. Each portion of the aperture serves as a source of secondary wavelets, observed at point P on the axis of the circular hole.

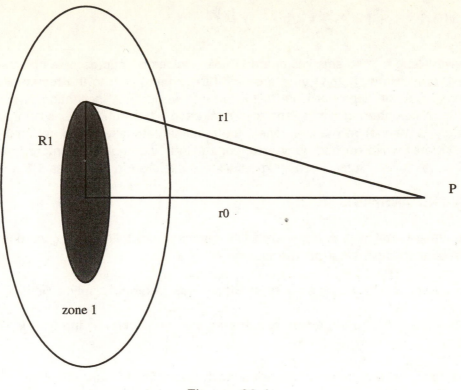

Figure 36-1

We choose the radius R_1 so that the path difference $(r_1 - r_0)$ between wavelets originating from a point a distance R_1 from the axis and those originating from the center is one-half wavelength. Thus wavelets coming from the points just above R_1 are "out of phase" with those coming from the center and destructively interfere with them. The circle of radius R_1 is called the first Fresnel zone (the shaded region of the figure). The second Fresnel zone would be an annular region bounded by radii R_1 and R_2. The radius R_2 is chosen so that the path difference for wavelets originating from that point $(r_2 - r_0)$ is one wavelength longer than those coming from the center. These relationships can be expressed mathematically using plane geometry.

$$r_1 = r_0 + \frac{\lambda}{2}; \quad \text{and} \quad r_2 = r_0 + 2\frac{\lambda}{2}; \quad \text{so that} \quad r_n = r_0 + n\frac{\lambda}{2}.$$

The radii R_1, R_2, ..., R_n, can be calculated by the Pythagorean theorem; leading to:

$$R_n^2 = nr_0\lambda + n^2\left(\frac{\lambda}{2}\right)^2$$

The area of the first Fresnel zone is then $A_1 = \pi R_1^2 = \pi[r_0\lambda + \lambda^2/4]$. The area of the nth Fresnel zone is obtained from the radii; $A_n = \pi[R_n^2 - R_{n-1}^2]$.

259

$$A_n = \pi[R_n{}^2 - R_{n-1}{}^2] = A_1 + \pi\left[2(n - 1)\left(\frac{\lambda}{2}\right)^2\right]$$

Because the wavelength is so small compared to dimensions of macroscopic objects, the Fresnel zones <u>all have approximately the same area</u>, and the contribution to the resultant electric field at the observation point is proportional to the area if we ignore "inclination effects", the contributions from neighboring zones tend to cancel each other out. Thus is an aperture presented an even number of Fresnel zones, the resultant intensity would be small (approximately zero) while an odd number of zones would contribute an intensity equal to that of a single Fresnel zone. Some interesting details are explored in examples 1,2, and 3.

QUESTIONS AND ANSWERS

Question. Suppose light was composed of tiny particles that followed trajectories like bullets. Would a diffraction pattern be formed?

Answer. No. They would produce sharp shadows, the "geometric optics" limit.

Question. What happens to diffraction patterns as the wavelength of the light is shortened, for a given slit size?

Answer. The patterns have maxima closely spaced about the forward direction or the patterns approach the "geometric optics" limit. The light behaves like bullets.

EXAMPLES AND SOLUTIONS

Example 1

Show that the quantity G defined to be

$$G \equiv \frac{1}{2} A_n - A_{n+1} + \frac{1}{2} A_{n+2}$$

is exactly zero using the result derived for the area of the $n\underline{th}$ Fresnel zone.

Solution:

From the definition of G we substitute the values for the zone areas;

$$G \equiv \frac{1}{2}\left\{A_1 + \pi\left[2(n - 1)\left(\frac{\lambda}{2}\right)^2\right]\right\} - \left\{A_1 + \pi\left[2(n)\left(\frac{\lambda}{2}\right)^2\right]\right\} + \frac{1}{2}\left\{A_1 + \pi\left[2(n + 1)\left(\frac{\lambda}{2}\right)^2\right]\right\}$$

The common area of the first Fresnel zone cancels out leaving

$$G \equiv \pi\left(\frac{\lambda}{2}\right)^2 \{n - 1 + n + 1 - 2n\} = 0.$$

260

Thus if the specific grouping designated by the factor G is used for adding up contributions from Fresnel zones, *exact cancellation occurs* if the resultant electric field vector from a zone is rigorously proportional to the area of the zone.

Example 2

Justify the statement that the intensity at the observation point P from an aperture with an infinite number of Fresnel zones is one-fourth that obtained from a single Fresnel zone.

Solution:

Based on the result of the previous example and the fact that the resultant electric field vectors from touching Fresnel zones are out of phase with each other, we write the total electric field contribution (E_T) at point P from the infinite number of Fresnel zones as:

$$E_T = \tfrac{1}{2} E_1 + \left[\tfrac{1}{2} E_1 - E_2 + \tfrac{1}{2} E_3 \right] + \left[\tfrac{1}{2} E_3 - E_4 + \tfrac{1}{2} E_5 \right] + ...$$

Now if the electric field vectors from each zone are proportional to the areas, then each of the groupings shown by the square brackets equals zero. This leaves

$$E_T = \tfrac{1}{2} E_1$$

Since the intensity is proportional to the square of the electric field, the intensity for an aperture with an infinite number of Fresnel zones is one-fourth that resulting from a single Fresnel zone. Put another way, if a wavefront is viewed at point P with no aperture at all and an intensity I_0 is obtained, then by placing a plane with a hole in it with area of one Fresnel zone (defined with respect to P), the intensity at P quadruples (i.e. $I_P = 4I_0$)!

Example 3

If the distance from a plane is $r_0 = 0.1$ m and $\lambda = 500$ nm, calculate the radius and the area of the first Fresnel zone.

Solution:

The radius of the first Fresnel zone is given by

$$R_1{}^2 = r_0\lambda + \left(\frac{\lambda}{2} \right)^2$$

where now $r_0 = 0.1$ m and $\lambda = 500$ nm. Substituting these values, we obtain:

$$R_1{}^2 = (0.1 \text{ m})(5 \times 10^{-7}\text{m}) + \left(\frac{5 \times 10^{-7}\text{m}}{2}\right)^2 \cong 5 \times 10^{-8}\text{m}^2$$

Therefore $R_1 = 2.236 \times 10^{-4}$ m = 0.224 mm, the second term making negligible contribution.

Example 4

For the single slit diffraction pattern, find the ratio of the intensity at the first maximum to that at the center.

Solution:

When δ is the total phase difference between the wavelets at the two extreme edges of the slit, the intensity is given by

$$I(\theta) = I_0 \frac{\sin^2(\delta/2)}{(\delta/2)^2}$$

where δ is a function of θ. If a is the slit width,

$$\delta(\theta) = \frac{2}{\lambda} (a \sin \theta).$$

(a) The first null occurs when $\delta/2 = \pi$ so $\sin \theta_1 = \lambda/a$. For small angles, $\theta_1 = \lambda/a$. There is a symmetrically placed dark fringe on the other side of the central bright spot so the angular width of the central bright spot is $\Delta\theta = 2\lambda/a$. The angular spacing between any other two dark fringes is λ/a.

(b) The first maximum occurs *near* $\delta/2$ equal to $3\pi/2$. The intensity at $\delta/2 = 3\pi/2$ is equal to:

$$I = I_0 \ 1/(3\pi/2)^2, \quad (\text{since } \sin 3\pi = 1.)$$

or

$$I = \frac{4}{9\pi^2} I_0 \cong \frac{1}{22} I_0$$

Example 5

A "replica" diffraction grating has 5276 lines per cm. Derive the intensity distribution due to such a grating and then obtain the positions of the first two maxima away from the central spot for the blue line in the hydrogen spectrum with wavelength of 486 nm.

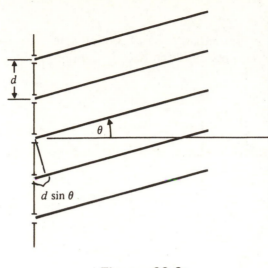

Figure 36-2a

Solution:

Referring to Fig. 36-2a, let d be the spacing between the slits. In cm, $d = (1/5276)$ cm = 1.895×10^{-4} cm. Consider N coherent sources, each one having a phase difference δ with the adjacent sources when viewed at an angle θ with respect to the normal. The path difference between two adjacent slits is $d \sin \theta$ so one would expect strong reinforcement when $d \sin \theta = n\lambda$. The central region $(n = 0)$ is a superposition of all the wavelengths (unless monochromatic light is used) in the incident beam so the various "orders" of the pattern correspond to $n = 1$, 2, 3, etc.

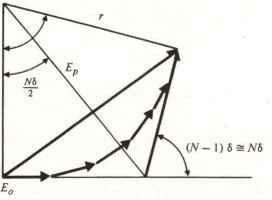

Figure 36-2b

To obtain the intensity distribution due to a finite number, N, of coherent sources, we

263

construct a phasor diagram where δ is now the common phase difference. (See Fig. 36-2b). The total phase difference is $(N - 1)\delta$ which we take to be $N\delta$ for N very large. The resultant electric field vector, E_p is

$$\frac{E_p}{2} = r \sin\left(\frac{N\delta}{2}\right)$$

where the radius of the construction circle, r, can be related to the electric field magnitude of a single source, E_0, and the common phase difference, δ, through:

$$r \sin\left(\frac{\delta}{2}\right) = \frac{E_0}{2}$$

This gives

$$E_p = \frac{E_0 \sin\left(\frac{N\delta}{2}\right)}{\sin\left(\frac{\delta}{2}\right)}$$

Squaring to obtain the intensity, we have

$$I = \frac{I_0 \sin^2\left(\frac{N\delta}{2}\right)}{\sin^2\left(\frac{\delta}{2}\right)}$$

where I_0 is the intensity due to one source. Notice that for very small δ, (the central bright spot), we can replace $\sin \delta/2$ by $\delta/2$ and in that limit the intensity, $I = N^2 I_0$! Without interference we would have expected a value of $N I_0$ but now it has been enhanced considerably since N can be a very large number.

To locate the maxima in the above pattern, we look for nulls in the denominator as this will give intense spikes of intensity with large dark regions in between. The sine function vanishes when the argument is an integral multiple of π so the condition for intensity maxima is:

$$\frac{\delta}{2} = n\pi \qquad n = 1, 2, 3, \text{ etc. (ignoring the central spot)}$$

Since $\delta = (2\pi/\lambda)d \sin \theta$, then $\sin \theta_n = n\lambda/d$ is the condition for maxima, just the condition guessed from the simple picture.

The coherence of these N sources is obtained by using a single light source to illuminate the series of slits. To evaluate the above expression numerically, we use $\lambda = 486$ nm and $d = 1895$ nm. Then $\sin \theta_n = n(0.2564)$ so $\theta_1 = 14.86°$, $\theta_2 = 30.86°$, $\theta_3 = 50.30°$, etc.

Example 6

An x-ray photon of energy E = 10 keV is incident on a NaCl single crystal where the lattice spacing d = 0.282 nm. What angle must the beam make with the surface of the crystal to satisfy the Bragg condition for constructive interference in first order?

Solution:

Since the Bragg condition is: $2d \sin \theta = n\lambda$; that condition for first order diffraction becomes: $2d \sin \theta_1 = \lambda$ or:

$$\sin \theta_1 = \frac{\lambda}{d}$$

The wavelength of the photon can be calculated from its known energy:

$$E_{photon} = h\nu = h \frac{c}{\lambda} = 10 \text{ keV}.$$

Using the values of h, c, and the conversion from eV to Joules, yields

$$\lambda = \frac{hc}{E} = \left(6.63 \times 10^{-34} \text{ J·s}\right) \frac{\left(3 \times 10^8 \text{ m/s}\right)}{\left(10^4 \text{ eV}\right)\left(1.6 \times 10^{-19} \text{ J/eV}\right)}$$

$$\lambda = 0.124 \text{ nm}.$$

Substitution of this value into the Bragg condition for first order gives:

$$\sin \theta_1 = \frac{0.124 \text{ nm}}{2(0.282 \text{ nm})} = 0.220 \qquad \text{or } \theta_1 = 12.7°.$$

Example 7

A binary star at a distance of 100 light years from earth is observed through a telescope using the blue line of hydrogen (486 nm). How large must the diameter of the lens be to see that there are two light sources present if the separation of the two component stars is 5 x 10-3 light years?

Solution:

The Rayleigh criterion is used here. Since the radius of the first dark ring is

$$R = 1.22 \left(\frac{\lambda}{D}\right) F$$

the angle subtended by this ring at the lens will be $\alpha_R = R/F$. For two light sources to be resolvable in the Rayleigh sense, their angular separation must exceed α_R for then the maximum of the second diffraction pattern will fall beyond the minimum in the first diffraction pattern. If the angular separation is less than α_R, the two sources will appear as a "blur" at the common center and be unresolvable.

In this case, the angular separation of the sources, α, is:

$$\alpha = \frac{\left(5 \times 10^{-3}\right)}{100} = 5 \times 10^{-5} \text{ rad.}$$

Thus if D is chosen such that $\alpha > \alpha_R$ the sources will be resolvable (we can tell that there are two of them). The condition is:

$$5 \times 10^{-5} > 1.22 \frac{\left(486 \times 10^{-9} \text{ m}\right)}{D}$$

Thus D must satisfy the inequality:

$$D > 1.19 \times 10^{-2} \text{ m}$$

QUIZ

1. Suppose an aperture when viewed by an observer at point P presents five Fresnel zones. Assume each zone contributes equal magnitude electric field vectors. (a) Calculate the intensity of the light seen using $I = KE_T^2$ where K is a constant and E_T is the total electric field at point P. (b) Calculate the resulting intensity at P if zones 2 and 4 are blocked off.

Answer: (a) $I = KE_1^2$ where E_1 is the electric field from zone 1; (b) $I = 9 KE_1^2$.

2. What energy photon would be required to satisfy the Bragg condition at $\theta = 15°$ *in the second order* for a lattice spacing of d = 0.282 nm?

Answer: E = 17 keV.

3. A circular aperture of radius $R = 2 \times 10^{-3}$ m is illuminated uniformly by light of wavelength 550 nm. How far from the aperture should one stand in order to see one and only one Fresnel zone?

Answer: 7.27 m

4. Light of wavelength 550 nm strikes a diffraction grating with mean spacing d. The first maximum away from the central bright spot occurs at $\theta = 15.96°$.
   (a) At what angle θ should one look to find the second maximum for light of wavelength 500 nm?
   (b) Compute the value of the grating spacing d.

Answer: (a) $\theta = 30°$; (b) $d = 2 \times 10^{-6}$ m.

37
RELATIVITY

OBJECTIVES

The objectives of this chapter are to:

Calculate the space and time differences between events in different inertial reference frames.

Calculate physical quantities such as momentum and energy relativistically.

Apply the work-energy relationship to the calculation of the velocity change of a particle.

Apply relativistic kinematics and dynamics to a variety of simple problems.

The relationship between space and time (and its implication) will be a substantial test for your physical intuition. Vectors with four (rather than three) components can be used to obtain all of the important results by essentially geometric reasoning.

REVIEW

While the equations developed in this chapter defy our intuition, they are nevertheless correct and the "special" theory of relativity, connected with Einstein's name, is not just another esoteric theory but its predictions have been reduced to engineering practice. The large particle accelerators built in Europe at CERN and in the United States at Fermi Lab are designed using relativistic mechanics and they work in the most minute detail.

The principle of relativity states that "the laws of physics are the same in every inertial frame of reference." The Galilean (or common sense) transformation from one inertial frame to another, namely,

$$x = x' + ut, \quad y' = y, \quad z' = z, \text{ and } t' = t,$$

insures that Newton's laws of motion are valid in all inertial frames. Unfortunately the additional requirement that the speed of light be the same for all observers conflicts with the Galilean transformation equations, which predict that the speed in some moving frame, c', would be related to the speed in the laboratory frame, c, by the equation

$$c' = c - u$$

where u is the relative velocity of the two frames (and can be positive or negative). To rectify this problem, the accepted concepts of space and time had to be modified so that a different set of transformation equations (the Lorentz transformation) could be obtained.

In particular, the idea that time intervals would be the same in two reference frames S and S' in a state of relative motion had to be modified. This can be seen from the famous thought experiment where a point source emits a spherical e.m. wave at $t = t' = 0$ when the origins of S and S' coincide. The relative motion of S and S' is along the x, x' axis. In S at a later time t, this wavefront is given by:

$$x^2 + y^2 + z^2 - c^2t^2 = 0$$

In S' at a later time t', we also must have a spherical wave front with:

$$x'^2 + y'^2 + z'^2 - c^2t'^2 = 0$$

Since the relative motion is along the x, x' direction, the transverse dimensions y and z are unaffected so $y = y'$ and $z = z'$ just as in the Galilean transformation. However now since the speed of light has the same value, c, in the two frames, setting $t' = t$ would imply that $x = x'$ which is incorrect since the two frames are in relative motion. Thus we can conclude that

$$x^2 - c^2t^2 = x'^2 - c^2t'^2$$

and that since $x \neq x'$, $t \neq t'$. The relativistically correct transformation must couple changes in x to changes in t in order to keep $x^2 - c^2t^2$ the same in all reference frames.

From the relativistically correct Lorentz transformation, it is apparent that two events that are simultaneous in one reference frame are not in another frame in relative motion with respect to the first frame.

The "time dilation" effect is a really new prediction of Einstein's relativity. The elapsed time t in frame S (which can be thought of as the lab frame) is longer than the elapsed time t' in the "moving" frame S' assuming the elapsed time in S' is measured by a clock at rest in S'. The "proper time" is the interval between two points occurring at the same space point.

The Lorentz contraction also is a new prediction of the Einstein relativity. In this effect, the longitudinal dimensions in the moving frame are reduced but the transverse dimensions are unaffected. This predication is easily obtained from the Lorentz transformation.

Two very interesting problems are encountered in Examples 6 and 7. In Example 6 the frequency shift called the Doppler effect is calculated for light, and in Example 7, the age difference for two twins is calculated when one twin takes a long, fast ride. The Doppler formula for light is much simpler than the corresponding formula for sound waves where the actual frequency shift depended on both the source and observer velocities. For light all that matters is the relative velocity of source and observer. The problem with the identical twins is a famous one and called the "Twin Paradox". There is still some doubt (not in the minds of most physicists) about this prediction as it has not been directly tested with living subjects. It has been tested and verified with atomic clocks.

The relativistic expressions for linear momentum, p, and energy, E, are different from the Newtonian Mechanics ones. Instead of $\vec{p} = m\vec{v}$, linear momentum is given by

$$\vec{p} = \frac{m\vec{v}}{\sqrt{1 - \left(v^2/c^2\right)}}$$

and the energy E, which is most comparable to our earlier "kinetic energy", since no potential energy is included is:

$$E = \frac{mc^2}{\sqrt{1 - \frac{v^2}{c^2}}}$$

The quantity m is called the rest mass and mc^2 is the rest energy, the value of E when $v = 0$. The quantity to be used for the kinetic energy is then $E - mc^2$ and when this quantity is expanded to lowest order in v/c we obtain that $E_k = E - mc^2 \cong (1/2)mv^2 + ...$. By manipulation of the above expression it is shown that:

$$E^2 = p^2c^2 + m^2c^4$$

This is a very important result. Since $(mc^2)^2$ would be the same in any reference frame, then $E^2 - (pc)^2$ must also have the same value in all frames (i.e. it is a relativistic invariant). Also in later chapters we will discuss particles that have no rest mass. For them, $m = 0$, so $E = pc$. The relationship between energy and momentum is illustrated in Example 8.

The constancy of the speed of light implies a very special law for the transformation of relative velocities. This expression, which can be obtained also from the Lorentz transformation is:

$$v' = \frac{v - u}{1 - \frac{uv}{c^2}}$$

Here v' is the velocity of a particle in frame S', v is its velocity in S, and u is the relative velocity of S and S'. This equation is capable of producing some surprising results. This is illustrated in Example 8.

The equivalence of mass and energy implied in the Einstein expression for E has been well tested in particle decay experiments. The amount of energy that can be obtained from conversion of a very small mass is intriguing but sacred conservation laws pertaining to heavy particles such as neutrons and protons limit the amount of energy that can be liberated by conversion of mass into energy. See Example 9.

PROBLEM-SOLVING STRATEGY

Try to identify the "events" in a given problem and assign to each event space and time coordinates. If that is possible, then the equations for the Lorentz transformation can be used for a solution. The invariance of the quantities such as: $x^2 - c^2t^2$ and $E^2 - (pc)^2$ can be used to solve many of the problems encountered in this chapter.

QUESTIONS AND ANSWERS

Question. Is the speed of sound the same in any reference frame? When you look up the speed of sound in a handbook, what reference frame is implied?

Answer. No. If you move toward a sound source, the speed of sound is higher than if you stand still with respect to the source. The speed quoted in a handbook is for the frame in which the medium of the sound--say air or water--is at rest.

Question. A photon has a speed of 0.95 c. Can its energy be increased by more than 5%? By more than 25%? By a factor of 500?

Answer. Yes to all questions. More energy increases the speed, but never in excess of the speed of light.

EXAMPLES AND SOLUTIONS

Example 1

Show that the constancy of the speed of light in all inertial reference systems implies that (a)

$$x^2 + y^2 + z^2 - c^2t^2 = x'^2 + y'^2 + z'^2 - c^2t'^2$$

and (b) if the relative motion of the two frames is along the xx' axis then

$$x^2 - c^2t^2 = x'^2 - c^2t'^2.$$

Solution:

(a) At time $t = t' = 0$ let the origins of the frames S and S' coincide. A light wave emitted at the origin at this time spreads out *spherically* in S and S'. In S the radius of the spherical wavefront is

$$R = (x^2 + y^2 + z^2)^{1/2} = ct$$

In S' the radius R' of the spherical wavefront is

$$R' = (x'^2 + y'^2 + z'^2)^{1/2} = ct'$$

Squaring these quantities we have $x^2 + y^2 + z^2 - c^2t^2 = 0$ and $x'^2 + y'^2 + z'^2 - c^2t'^2 = 0$ so that

$$x^2 + y^2 + z^2 - c^2t^2 = x'^2 + y'^2 + z'^2 - c^2t'^2 = 0$$

Since this quantity is the same in any two inertial frames S and S' it is called a relativistic *invariant* and is very useful in calculations.

(b) For a relative motion parallel to the xx' axis, the dimensions perpendicular to the motion are unaffected so that

$$y = y' \quad \text{and} \quad z = z'$$

and the above relativistic invariant reduces to

$$x'^2 - c^2t'^2 = x^2 - c^2t^2$$

Example 2

Imagine a light clock in a frame of reference (labeled M) moving with respect to the "laboratory" frame (labeled L) with velocity V. The first event occurs at $t_M = t_L = 0$ at the space points $x_M = x_L = 0$, when the two origins coincide and a pulse of light leaves the first mirror and travels along the Y_M axis between two parallel mirrors. The second event occurs when the pulse arrives at the second mirror, still at $x_M = 0$ but at a later time $t_M/2$. The third event occurs when the pulse arrives back at the original mirror, still at $x_M = 0$ but now at a later time t_M. (a) Give a sketch of the events in the LAB frame and find the relationship between the time intervals in the moving frame (t_M) and and in the lab frame (t_L). (b) Suppose V = 0.6c.   If the light clock is designed so that the interval between arrivals at the same mirror is equal to 1 second, find the corresponding interval in the lab frame.

Solution

(a) Referring to Figure 37-1, the three events in the lab frame appear as an equilateral triangle.

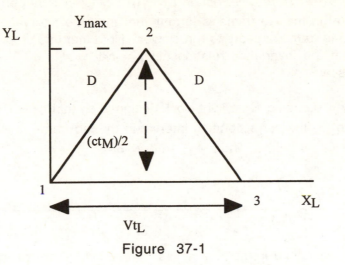

Figure 37-1

The time interval in the moving frame is given by:

$$t_M = \frac{2Y_{max}}{c}$$

where Y_{max} is the vertical distance between the two mirrors and the __same__ in the two frames. In the lab frame, the corresponding time interval is:

$$t_L = \frac{2D}{c}$$

where D is the hypotenuse of the right triangle of height Y_{max} and base $Vt_L/2$. From the right triangle, we have

$$D^2 = \left(Y_{max}\right)^2 + \left(\frac{Vt_L}{2}\right)^2$$

Substituting for D and Y_{max} leads to:

$$\left(\frac{ct_L}{2}\right)^2 = \left(\frac{ct_M}{2}\right)^2 + \left(\frac{Vt_L}{2}\right)^2$$

Re-arranging this expression and solving for the time interval in the moving frame:

$$t_M^2 = \left(1 - \beta^2\right) t_L^2 = \frac{t_L^2}{\gamma^2}$$

By taking the positive square-root we obtain $t_L = \gamma t_M$. Since γ is larger than unity, the apparent time interval in the lab frame is <u>longer</u> that the time interval in the moving frame; the clock in the moving frame appears to run slower. Note that this problem can also be solved (and easily) by using the invariance relationship $x^2 - c^2 t^2 = x'^2 - c^2 t'^2$. Substitute the values $x' = 0$, $x = Vt$, and solve for t'.

(b) Numerically for $V = 0.6c$ and a the light clock designed so that $t_M = 1$ second, we use the above result to solve for the corresponding interval in the lab frame, t_L. We are given that $\beta = 0.6$ so that $(1 - \beta^2) = 0.64$. Then $\gamma = 1/0.8 = 5/4$.

$$t_L = \gamma t_M = (1.25)(1 \text{ s}) = 1.25 \text{ s}.$$

Example 3

A a stick of length L_0 rests in a frame that moves with speed V with respect to the lab frame. The coordinates of the stick are $x_{1M} = 0$ and $x_{2M} = -L_0$. Thus $x_{1M} - x_{2M} = L_0$, the proper length of the stick. An observer in the lab frame measures the length of the stick, in the lab frame by measuring the time difference between two events; the first event occurs when the leading edge of the stick passes the origin of the lab frame (say at $t_L = t_M = 0$) and the second event when the trailing edge of the stick passes the origin, a time interval Δt_L later. What value does this observer obtain for the length of the stick?

Solution

The "apparent length" of the stick is then $V \Delta t_l = L$. The Lorentz transform equations can be applied to this measurement problem.

$$t_M = \gamma \left[t_L - \left(\frac{V}{c^2} \right) x_L \right] \text{ and } x_M = \gamma \left(x_L - V t_L \right)$$

and the inverse relationship:

$$t_L = \gamma \left[t_M + \left(\frac{V}{c^2} \right) x_M \right] \text{ and } x_L = \gamma \left(x_M - V t_M \right)$$

Since $x_L = 0$ for both measurements, $x_M = -V t_M = -L_0$. Therefore for Δt_L,

$$\Delta t_L = \gamma \left[\frac{L_0}{V} + \left(\frac{V}{c^2} \right)(-L_0) \right] = \gamma \frac{L_0}{V} \left(1 - \beta^2 \right)$$

Since $V\Delta t_L = L_L = \gamma L_0(1 - \beta^2) = L_0\sqrt{(1 - \beta^2)}$, the length measured in the lab of the stick is shorter than the actual length. It appears that the stick has shrunk in the dimension of the motion, the other dimensions remaining the same.

Example 4

Two events occur simultaneously in frame S where they are separated by 1.28×10^7 m (the diameter of the earth).
 (a) Is it possible to find a reference frame S' where these two events occur at the same space point ?
 (b) In a reference frame S' where the two events are separated by 2.56×10^7 m, calculate the time interval between the events and
 (c) the relative velocity of the frames S and S'.

Solution:

(a) It is *not* possible to find a frame where these two events occur at the same point (or even closer together). This can be seen by using the invariant quantity

$$(\Delta x')^2 - c^2(\Delta t')^2 = (\Delta x)^2 - c^2(\Delta t)^2$$

In S, $\Delta t = 0$ so that

$$(\Delta x')^2 - c^2(\Delta t')^2 = (\Delta x)^2 = \text{a positive real number}$$

If $\Delta x' = 0$ then $\Delta t'$ cannot be a real number. For $\Delta t'$ to be real we must have $\Delta x' > \Delta x$ in this problem.

(b) Writing the invariant for the simultaneous events in S

$$(\Delta x')^2 - c^2(\Delta t')^2 = (\Delta x)^2$$

and substituting

$$\Delta x = 1.28 \times 10^7 \text{ m}$$

$$\Delta x' = 2.56 \times 10^7 \text{ m}$$

we have

$$(\Delta t')^2 = \frac{(2.56 \times 10^7 \text{ m})^2 - (1.28 \times 10^7 \text{ m})^2}{(3 \times 10^8 \text{ m/s})^2}$$

$$\Delta t' = 7.39 \times 10^{-2} \text{ s}$$

275

The relative velocity of the two frames can be obtained from the Lorentz transformation equation

$$x_2' - x_1' = \frac{(x_2 - x_1) - u(t_2 - t_1)}{\sqrt{1 - u^2/c^2}}$$

with $t_2 - t_1 = 0$. Since $x_2' - x_1' = 2(x_2 - x_1)$ we have

$$\sqrt{1 - \frac{u^2}{c^2}} = \frac{1}{2}$$

Squaring results in

$$1 - \frac{u^2}{c^2} = \frac{1}{4}$$

Solving for the velocity, we have

$$u = \frac{\sqrt{3}}{2}\, c = 0.866\, c = 2.60 \times 10^8 \text{ m/s}$$

Example 5

An electron is accelerated from rest to a velocity $v = 0.9\, c$ by means of a potential difference ΔV. Calculate ΔV using 0.511 MeV for the rest energy of the electron.

Solution:

The energy of the electron is

$$E = \frac{mc^2}{\sqrt{1 - \frac{v^2}{c^2}}}$$

$$E = \frac{0.511 \text{ MeV}}{\sqrt{1 - (0.9)^2}} = 1.172 \text{ MeV}$$

The increase in kinetic energy is equal to the work done on the particle by the accelerating potential

$$e\Delta V = \Delta K$$

$$= E - mc^2$$

$$= 1.172 \text{ MeV} - 0.511 \text{ MeV} = 0.661 \text{ MeV}$$

Solving for ΔV we have

$$\Delta V = 6.61 \times 10^5 \text{ volts} = 661,000 \text{ volts}$$

Example 6

If a source of electromagnetic radiation at rest in S' produces regular pulses with spacing τ', what is the separation of these pulses in S?

Solution:

Let the first pulse be emitted at $t = t' = 0$ when the origins coincide. If the second pulse is emitted at $t' = \tau'$, then the time interval t in the frame S is equal to:

$$t = \frac{\tau'}{\sqrt{1 - u^2/c^2}}$$

This pulse is emitted at a value of x equal to ut and so must travel back to the observer at the origin of S taking an additional time equal to x/c. Therefore the time between the two pulses, τ, according to an observer at the origin of S is

$$\tau = t + \frac{x}{c} = t + \frac{ut}{c} = \left(1 + \frac{u}{c}\right)t$$

or

$$\tau = \frac{(1 + u/c)}{\sqrt{1 - u^2/c^2}} \tau' = \frac{\sqrt{1 + u/c}}{\sqrt{1 - u/c}} \tau'$$

If we regard τ and τ' as the respective periods of the sources, then the frequencies f and f' would be given by:

$$f = \frac{\sqrt{1 - u/c}}{\sqrt{1 + u/c}} f' \qquad \text{(Doppler Shift)}$$

If the source (in S') is moving away from the observer (in S), then f is less than f'. The frequency of a receding train's whistle is lowered. If the source is moving toward the observer, then u is negative and f > f'.

Known spectral lines from hydrogen and helium seen in the radiation reaching the earth from distant galaxies are shifted in wavelength toward the red part of the visible spectrum. This "red shift" combined with the previous expression for the Doppler shift is evidence that these galaxies are receding from us and an important component of the "Big Bang" theory of the universe.

Example 7

If a "moving clock" runs slower, what will the age difference be between two twins if one stays on the earth while the second makes a round trip to a point in space ten light years from the earth at a speed of 0.95c?

Solution:

The distance, D, traveled is equal to the speed of light multiplied by twenty years. The time taken for this trip according to the twin on earth, T, is:

$$T = \frac{c(20 \text{ years})}{0.95 \text{ c}} = 21.05 \text{ years}$$

The time elapsed on the "moving clock", T', is related to T by the time dilation formula:

$$T = \frac{T'}{\sqrt{1 - u^2/c^2}}$$

Numerically, we have $[1-u^2/c^2]^{1/2} = 0.312$ so T' = 6.57 years. Therefore if the twins were 30 years old when the separation occurred, the twin left on earth is 51 years old at the reunion whereas the space traveller is 36.6 years old.

Example 8

Two electrons traveling in the same direction have energies of 1 Mev and 2 Mev respectively as seen from frame S. Find the velocity of each of these electrons in S and then find the velocity of the most energetic one relative to that of the least energetic one. Use 0.51 Mev for the electron rest energy.

Solution:

Since the energy is

$$E = \frac{mc^2}{\sqrt{1 - \frac{u^2}{c^2}}}$$

where mc^2 is the rest energy, for electron 1 we have:

$$1 \text{ MeV} = \frac{(0.51 \text{ MeV})}{\sqrt{1 - u_1^2/c^2}}$$

Solving numerically for u_1 we have:

$$u_1/c = 0.8602.$$

278

For electron 2,

$$2 \text{ MeV} = \frac{(0.51 \text{ MeV})}{\sqrt{1 - u_2^2/c^2}}$$

so that

$$u_2/c = 0.9669$$

Since the electrons are travelling in the same direction we have chosen both positive signs for the u's.

To find the velocity of electron 2 with respect to electron 1, we must use the Einstein addition law for relative velocities. *The relative velocity is not* $u_2 - u_1 = c(.9669 - .8602) = 0.1067 \; c.$

The equation given for the addition of relative velocities in the text is:

$$v' = \frac{v - u}{1 - \frac{u\,v}{c^2}}$$

where we consider the frame S' to be the frame where electron 1 is at rest. Thus we interpret the symbols as:

v = velocity of electron 2 in S = 0.9669 c
u = velocity of S' with respect to S = u_1 = .8602 c
v' = unknown velocity of electron 2 in S'

$$v' = \frac{0.9669 \; c - 0.8602 \; c}{1 - (0.8602)(0.9669)} = 0.636 \; c$$

Since electron 1 is at rest in S', this is the relative velocity of electron 2 with respect to 1. Note that it is more than a factor six larger than the classically expected result.

Example 9

When the neutron spontaneously decays into a proton, an electron and a neutrino (which is massless), the decay products are observed to have a total kinetic energy of 1.25×10^{-13} J. If the proton mass (M_p) is 1.673×10^{-27} kg and the electron mass (M_e) is 9.110×10^{-31} kg, how large is the neutron mass (M_N)?

Solution:

We equate the rest energy of the neutron to the total energy of the by-products. This energy of the by-products is equal to the sum of the rest energies of the proton and electron (the neutrino is massless) and the kinetic energy of the by-products.

$$M_N c^2 = M_p c^2 + M_e c^2 + E_k$$

or

$$M_N = M_p + M_e + \frac{E_k}{c^2}$$

$$= (1.673 \times 10^{-27} + 9 \times 10^{-31} + 1.39 \times 10^{-30}) \text{ kg}$$

$$= (1673 + 2.3) \times 10^{-30} \text{ kg}$$

$$= 1.675 \times 10^{-27} \text{ kg}$$

Because $M_N > M_p$, the proton cannot spontaneously decay into a neutron (plus a position and neutrino).

QUIZ

1. Two events in frame S occur at the same space point with a time difference of 4 s. In a frame S' moving relative to S with speed u, the separation in time of the two events is 5 s. Calculate
    (a) the separation between the positions of the two events in S'
    (b) the relative velocity of the two frames.

Answer: (a) 9×10^8 m, (b) u $= \pm 0.6$ c

2. An electron initially at rest is accelerated through a potential difference of 10^6 V. Calculate the final velocity of the electron.

Answer: $(v/c) = 0.9988$ or v $= 2.997 \times 10^8$ m/s

3. The rest mass of the proton is $(1836)m_0$ where m_0 is the electron rest mass. Assuming the above relationship is exact (it isn't), calculate the ratio of the electron velocity to the speed of light that would make the electron "mass" as large as the proton rest mass.

Answer: $v/c = 0.99999985$

4. The kinetic energy of a beam of electrons is 200 keV. Calculate the mass of the the electrons in the beam, in units of the electron rest mass, m_0 .

Answer: m $= 1.391\ m_0$

5. In the CERN storage ring, μ mesons whose decay half-life at rest is 2.200 µs, are accelerated to a speed (v) near the speed of light. In that reference frame, their half-life is observed to be 26.2 µs. Calculate the speed of that reference frame (the μ mesons).

Answer: v = 0.996 c.

38

QUANTUM PHYSICS I:
PHOTONS, ELECTRONS, AND ATOMS

OBJECTIVES

In this chapter your objectives are to:

Calculate the threshold wavelength for light to eject photoelectrons from a surface with a given work function.

Calculate the stopping potential for a given wavelength for photoelectrons.

Calculate the frequencies of absorbed and emitted light given a set of energy levels.

Calculate the wavelength change of an x-ray that is scattered from an electron.

Recognize the *nucleus* as a small core (size about 10^{-14} m) inside the larger atom (size about 10^{-10} m) containing all the positive charge and most of the mass of the atom.

Calculate the energy levels of the Bohr atom.

Electromagnetic radiation, previously explained by means of a wave picture, is endowed with some particle-like features, just as particles will be endowed with wave-like features in the following chapters.

REVIEW

Electromagnetic waves in the radiofrequency range (about 10^8 Hz) can be understood in terms of classical electromagnetism but higher frequency radiation (about 10^{15} Hz), light, produces several puzzles particularly in its interaction with matter which can not be understood in the framework of Maxwell's equations and Newton's laws. Maxwell's equations predict that accelerating charges radiate electromagnetic waves. In the late 1800's it was speculated that light came from a motion of electric charge in individual atoms. The experiments of that era were too crude to confirm or deny this. Some of the puzzles were:

(1) Each element emits a characteristic spectrum when it is heated in a flame or excited electrically (like neon in a glow tube). This spectrum does not contain all possible wavelengths but only certain well defined ones. This suggests that emitted light is related to the characteristics and internal structure of an atom -- an idea that cannot be explained with classical theories.

(2) The photoelectric effect could not be explained classically. For thermionic emission of electrons from a surface where the kinetic energy needed by an electron to escape from the surface is supplied by thermal motions, some electrons escape even from cold surfaces. If light is directed onto the surface, and the frequency of the light is wrong, then no electrons are ejected even though the source intensity is high. If the frequency is properly chosen, electrons can be ejected but they have a maximum kinetic energy *that does not depend on the source intensity*.

(3) The problems with light extend to higher frequencies (10^{19} Hz) where electromagnetic radiations are called x-rays. The problem here was that the scattered x-rays frequently had a longer wave- length than the original x-ray (which the text likens to a light wave suffering a color change upon reflection).

Although this doesn't exhaust the list of problems, it does suffice to indicate that classical theories didn't have the necessary components for a complete solution. The explanations that were offered on the basis of classical theory were many times more puzzling that the original puzzles.

THE PHOTOELECTRIC EFFECT

Einstein suggested that the photoelectric effect could be understood if light had a particle-like character. These light quanta were called photons and their energy was directly related to the frequency of the light wave, f, through

$$E = hf$$

with h = 6.626 x 10^{-34} J·s being Planck's constant. To eject an electron from a metal, the photon energy would have to be larger than the potential energy holding the electron in the metal (called the work function, ϕ). An electron would not be ejected if it had to wait for several quanta to come along and provide it with enough energy. This explained why even intense sources of light would not eject electrons if the light frequency were too low. Further, the ejected electrons would leave the surface with a kinetic energy

$$\frac{1}{2} mv^2 = hf - \phi$$

If a potential difference, V_0, of suitable polarity was applied, with $eV_0 = 5mv^2$ then the electrons could be stopped and no current flow would result. Examples 2 and 3 illustrate aspects of the photoelectric effect.

THE NUCLEAR ATOM

Rutherford scattering of alpha particles (the nuclei of helium atoms) demonstrated that the positive charge and nearly all the mass of an atom were concentrated in a region of space small compared to that occupied by an atom. The dense charged region is called the nucleus.

The nuclear diameter is of order 10^{-14} m while the atomic diameters are typically 10^{-10} m making the fraction of the atomic volume occupied by the nucleus about 10^{-12}!

THE BOHR MODEL

Although Rutherford had suggested a model of the atom that had the positive charge and most of the mass concentrated at the center of the atom with electrons orbiting the nucleus -- much like our solar system -- the deficiencies of this model prevented it from being accepted. The main problem stemmed from the prediction of Maxwell's equations that an accelerating charge must radiate energy at the frequency of the circular revolution. The orbit would collapse due to the loss in energy and the frequency spectrum from an atom would be continuous rather than a sharp line spectrum.

Bohr's postulates are used to obtain the energy levels of a one electron atom. The critical postulate is that the angular momentum must be an integral multiple of $(h/2\pi)$. Incorporation of that one idea into an otherwise classical framework produces a model of the atom with quantized negative energy levels (the zero of potential energy was chosen to coincide with infinite separation between the two charges) which forms the basis of understanding for the observed sharp line spectrum of atoms. Example 6 compares the H atom allowed energies with those of a helium atom with one electron missing, an ionized He atom. Example 7 repeats the Bohr model calculation for a bound system of an electron and position that forms an atom called positronium.

The line spectra emitted by the elements were explained by the Bohr model of the atom. In this model, picture all of the positive charge as well as most of the mass concentrated in the center or the nucleus and the electron orbiting about this nucleus much like the planets orbit about the sun. The postulates for this model of the hydrogen atom are:

(1) The electron (charge -e) travels about the proton (charge +e) in a circular orbit due to the Coulomb attraction between them.

(2) Not all orbits are allowed. The allowed ones are those for which the angular momentum of the electron about the proton is $nh/2\pi$ where n is an integer 1,2,3 etc. and $h = 6.626 \times 10^{-34}$ J·s is Planck's constant. The quantity $h/2\pi$ occurs so frequently that it is designated by the symbol $\hbar = h/2\pi$ and called "h-bar".

(3) While in one of these allowed orbits, the constantly accelerating electron in its circular orbit does not radiate energy, which would cause it to undergo a "death spiral" into the nucleus. The energy characteristic of the allowed orbits, E_n, is constant in time.

(4) Radiation is emitted by the atom *only* when the electron makes a transition from one allowed orbit, E_n, to another E_n'. The frequency of the emitted radiation is:

$$hf = E_n - E_n'$$

With these postulates, the radii of the allowed orbits, r_n and the allowed energy levels, E_n, are given by

$$r_n = n^2 r_0$$

$$E_n = -(13.6 \text{ eV})/n^2$$

where r_0, numerically equal to 0.53×10^{-10} m, is given in terms of fundamental constants.

$$r_0 = \frac{4\pi\varepsilon_0 \hbar^2}{me^2} = 0.53 \times 10^{-10} \text{ m}$$

Application of the above formulas gives the positions of the spectral lines of hydrogen satisfactorily and also indicates many of the gross features of atomic spectra in general. Beyond this however, the Bohr model must be replaced by a model obtained from quantum mechanics as it suffers many shortcomings of its own.

X-RAY PRODUCTION AND SCATTERING

Just as shining light on a metallic surface (in vacuum) could eject electrons, bombarding a surface with electrons can produce electromagnetic radiation. In particular, very short wavelength, high frequency radiation called x-rays can be produced in this manner. Interaction of such x-rays with atoms can lead to ejection of a tightly bound inner core electron. The filling of this vacancy by an outer electron leads to a characteristic line spectrum (that can be used to identify elements) like those observed for simple atoms such as H or He.

The wavelength shift in scattered x-radiation, called the Compton effect, was satisfactorily explained by using the particle-like nature of radiation (the photon picture) and applying the conservation of linear momentum and kinetic energy, (i.e. assume the collision is elastic) to the "collision" of an electron, at rest initially, with a photon. Since the photon momentum, p, is equal to h/λ, the momentum given to the electron reduces that of the photon -- increasing its wavelength to the value λ'. The change in wavelength is also a function of the angle θ between the incident x-ray and the emerging one. Explicitly we have:

$$\lambda' - \lambda = \frac{h}{mc}(1 - \cos\theta)$$

with m the electron mass. Applications of this formula are made in Examples 9 and 10.

CONTINUOUS SPECTRA

In contrast to the sharp spectral lines emitted by gases, radiation from liquid and solid surfaces is in the form of a continuous spectrum. This radiation has been termed "blackbody radiation" and has been extensively studied. We encountered it first in Thermodynamics when considering heat transfer mechanisms. Blackbody radiation does not consist of radiation at a single frequency or wavelength but instead is distributed over all wavelengths. The spectrum (or spectral emittance) was first discovered by Plank. The distribution of wavelengths is given by:

$$I(\lambda) = \frac{2\pi hc^2}{\lambda^5\left(e^{hc/\lambda kT} - 1\right)}$$

This spectrum has a peak at $\lambda_m T = 2.90 \times 10^{-3}$ m·K. See Example 11. The total power per unit area (I) is found by integrating the spectral emittance $I(\lambda)$ over all wavelengths. This produces the famous result

$$I = \int_0^\infty I(\lambda) \; d\lambda = \sigma T^4$$

where σ is the Steffan constant encountered previously.

QUESTIONS AND ANSWERS

Question. Suppose a photon in the green range of the spectrum can eject an electron from a metal but a photon in the red range of the spectrum can't. Could a photon in the blue range eject an electron from the same metal?

Answer. Yes. Photons in the blue range are more energetic than photons in the green range because the energy of a photon is proportional to its frequency and blue light has a higher frequency than green light.

Question. In the Compton Effect, why is there zero wavelength shift for forward scattering $(\theta = 0°)$?

Answer. In forward scattering no energy or momentum is transferred to the electron; if they were, conservation of momentum and energy could not be simultaneously satisfied.

EXAMPLES AND SOLUTIONS

Example 1

Find the average number of quanta that strike the earth per m2 per second from the sun. Use for data: the value of the solar constant S = 1.4 kW/m2; the Sun surface temperature 6000 K; and the Wien displacement law $\lambda_m T = 2.9 \times 10^{-3}$ m·K.

Solution

The mean wavelength, from the Wien displacement law, is equal to:

$$\lambda_m = \frac{2.9 \times 10^{-3} \; m \cdot K}{T_{sun}} = \frac{2.9 \times 10^{-3} \; m \cdot K}{6000 \; K} = 4.8 \times 10^{-7} \; m$$

The energy of a quantum with this wavelength, e = hf is:

$$e = hf = \frac{hc}{\lambda} = \frac{\left(6.63 \ x \ 10^{-34} \ J \cdot s\right)\left(3 \ x \ 10^{8} \ m/s\right)}{\left(4.8 \ x \ 10^{-7} \ m\right)} = 4.1 \ x \ 10^{-19} \ J$$

If we divide the given solar constant by the mean energy of a quantum, we obtain the number (n) of quanta per second per m2 striking the cross-section of the earth (area $\pi R_e{}^2$):

$$n = \frac{S}{e} = \frac{1.4 \ x \ 10^{3} \ J/s \cdot m^{2}}{4.1 \ x \ 10^{-19} \ J} = 3.4 \ x \ 10^{21} \ quanta/s \cdot m^{2}$$

Example 2

The work functions of several metals are listed below. Which metals yield photoelectrons when bombarded with light of wavelength 500 nm? For those surfaces where photoemission occurs with the above light source, calculate the stopping potential in volts.

Solution:

Metal	ϕ (in eV)
W	4.5
Ag	4.8
Cs	1.8
Cs on W	1.36

Since only one light source is involved, it is best to calculate the energy of a photon in eV for this source.

$$E = hf = h\frac{c}{\lambda} = \left(6.626 \ x \ 10^{-34} \ J \cdot s\right)\frac{\left(3 \ x \ 10^{8} \ m/s\right)}{\left(5 \ x \ 10^{-7} \ m\right)}$$

$$= 3.97 \ x \ 10^{-19} \ J = 2.48 \ eV.$$

Thus the visible light source will not produce electrons if it shines on W or Ag. The stopping potential for Cs is 2.48 eV – 1.8 eV or 0.68 V. For Cs on W, the stopping potential is 1.12 Volts.

Example 3

For the metals tungsten (W) and silver (Ag) in Example 2, calculate the threshold wavelength which would just start producing photoelectrons.

Solution:

The kinetic energy of the photoelectrons is given by

$$hf = h\frac{c}{\lambda} = \frac{1}{2}mv^2 + \phi$$

When the wavelength is the threshold wavelength, the photoelectrons have *zero* kinetic energy. Thus we have

$$h\frac{c}{\lambda_t} = \phi$$

Solving for λ_t results in

$$\lambda_t = \frac{hc}{\phi}$$

The work functions for W and Ag are 4.5 eV and 4.8 eV respectively. For W this yields

$$\lambda_t = \frac{(6.626 \times 10^{-34}\ J\cdot s)(3 \times 10^8\ m/s)}{(1.6 \times 10^{-19}\ C)(4.5\ eV)} = 276\ nm.$$

For Ag, using the previous value and the ratio of the work functions

$$\lambda_t = \frac{(4.5\ eV)}{(4.8\ eV)}\ 276\ nm = 259\ nm.$$

These wavelengths not in the "visible" region of the spectrum.

Example 4

The stopping potential for photoelectrons ejected from a surface by 375 nm photons is 1.870 volts. Calculate the stopping potential if 600 nm photons are used.

Solution:

The stopping potential is related to the wavelength and the work function by

$$eV_0 = h\frac{c}{\lambda} - \phi$$

For a second wavelength λ' the stopping potential would be V_0' where

$$eV_0' = h\frac{c}{\lambda'} - \phi$$

Subtracting these two equations eliminates ϕ

$$eV_0 - eV_0' = h\frac{c}{\lambda} - h\frac{c}{\lambda'}$$

Solving for the unknown stopping potential results in

$$V_0' = V_0 - \frac{hc}{e}\left[\frac{1}{\lambda} - \frac{1}{\lambda'}\right]$$

$$V_0' = 1.870 \text{ V} - \frac{(6.626 \times 10^{-34} \text{ J·s})(3 \times 10^8 \text{ m/s})}{(1.6 \times 10^{-19} \text{ C})}\left[\frac{10^9}{375 \text{ m}} - \frac{10^9}{600 \text{ m}}\right]$$

$$= 1.870 \text{ V} - 1.242 \text{ V}$$

$$= 0.628 \text{ V}.$$

Example 5

An atom has in addition to the ground state energy E_0 (taken to be zero) levels $E_1 = 10.20$ eV, $E_2 = 12.09$ eV, and $E_3 = 12.75$ eV. If the atom is excited from its ground state to the state with energy 12.75 eV, calculate the wavelengths of the lines that might exist in the spectrum of this atom.

Solution:

The *possible* frequencies satisfy the equation

$$hf = E_n - E_m > 0.$$

For transitions out of the state E_3, the atom could go to level E_2, E_1, or E_0, giving frequencies

$$hf_1 = 12.75 \text{ eV} - 12.09 \text{ eV} = \frac{hc}{\lambda_1}$$

$$\lambda_1 = \frac{(6.626 \times 10^{-34} \text{ J·s})(3 \times 10^8 \text{ m/s})}{(0.64 \text{ eV})(1.6 \times 10^{-19} \text{ C})} = 1.88 \times 10^{-6} \text{ m}$$

$$hf_2 = 12.75 \text{ eV} - 10.20 \text{ eV} = \frac{hc}{\lambda_2}$$

$$\lambda_2 = 4.87 \times 10^{-7} \text{ m}.$$

$$hf_3 = 12.75 \text{ eV} - 0 = \frac{hc}{\lambda_3}$$

$$\lambda_3 = 9.73 \times 10^{-8} \text{ m.}$$

Furthermore, from state E_2 the atom could go to states E_1 and E_0:

$$hf_4 = 12.09 \text{ eV} - 10.20 \text{ eV} = \frac{hc}{\lambda_4}$$

$$\lambda_4 = 6.57 \times 10^{-7} \text{ m}$$

and

$$hf_5 = 12.09 \text{ eV} - 0 = \frac{hc}{\lambda_5}$$

$$\lambda_5 = 1.02 \times 10^{-7} \text{ m}$$

Finally from state E_1 the atom could make a transition to the ground state:

$$hf_6 = 10.20 \text{ eV} - 0 = \frac{hc}{\lambda_6}$$

$$\lambda_6 = 1.22 \times 10^{-7} \text{ m.}$$

Example 6

Compare the allowed energy levels for the Bohr hydrogen atom with those of a singly ionized helium atom (He+).

Solution

For *hydrogen*, the allowed energy levels in the Bohr model are:

$$\left(E_n\right)_{hydrogen} = -\frac{hcR}{n^2} = -\frac{me^4}{2\left(4\pi\varepsilon_0\right)^2 \hbar^2 n^2} = -\frac{13.6 \, eV}{n^2}$$

For singly ionized helium, there is only one electron, like hydrogen, but there are two protons (and one or two neutrons) in the nucleus so the nuclear charge is +2e. If the nuclear charge was Ze, the factor e^4 in the numerator, coming from the Coulomb interaction (squared) between the electron and the nucleus, would be replaced by $(Ze^2)^2 = Z^2 e^4$. Changing the numerator in the previous expression from e^2 to $4e^2$ (since Z = 2) leads to:

$$\left(E_n\right)_{He^+} = -\frac{4me^4}{2\left(4\pi\varepsilon_0\right)^2\hbar^2 n^2} = -\frac{54.4\,eV}{n^2}$$

Example 7

The particle called the positron has the same mass as the electron but opposite electrical charge. A bound state of an electron and a positron is an atom called positronium. Treating this as a Bohr atom, calculate its energy levels.

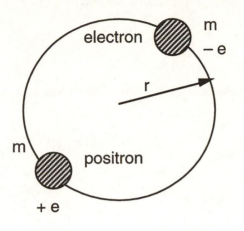

Figure 38-1

Solution:

As a reminder, the electron energy levels in the Bohr atom for Z = 1 are

$$E_n = -\frac{1}{2}\frac{me^4}{\left(4\pi\varepsilon_0\right)^2\hbar^2}\frac{1}{n^2} = -\frac{Rch}{n^2}$$

where the product Rch, R being the Rydberg constant, has the numerical value 13.6 eV. Referring to Fig. 38-1, we see that the potential energy of this bound system (zero at r = ∞) is:

$$E_p = -\frac{1}{4\pi\varepsilon_0}\left(\frac{e^2}{2\,r}\right)$$

where 2r is the separation. The kinetic energy is $E_k = (1/2)mv^2 + (1/2)mv^2$ so we have

$$E = E_k + E_p = mv^2 - \frac{1}{4\pi\varepsilon_0}\left(\frac{e^2}{2\,r}\right)$$

The *total* angular momentum is the quantized entity:

$$mvr + mvr = n\hbar = 2mvr.$$

The force on the electron that causes it to stay on the circle and provides the centripetal acceleration is the Coulomb force:

$$\frac{1}{4\pi\varepsilon_0} \frac{e^2}{(2r)^2} = \frac{mv^2}{r}$$

Using this last equation along with the energy equation, we find

$$E = -\frac{e^2}{4\pi\varepsilon_0} \frac{1}{(4r)}$$

By squaring the angular momentum equation and comparing with the above expression for v^2, we obtain:

$$r = \frac{n^2\hbar^2 4\pi\varepsilon_0}{me^2}$$

The energy is then:

$$E_n = \frac{me^4}{2\left(4\pi\varepsilon_0\right)^2\hbar^2} \frac{1}{n^2} \frac{1}{2} = -\frac{Rch}{2} \frac{1}{n^2}$$

Thus the "effective" Rydberg constant for positronium is half as large as that for hydrogen resulting in a binding energy for positronium of 6.8 eV.

Example 8

In a Bohr atom, (a) calculate the frequency of radiation emitted when the atom makes a transition from a state with quantum number n+1 to a state with quantum number n; (b) the frequency associated with rotation in the nth Bohr orbit; and (c) show that in the limit of very large quantum number n that the answers to parts (a) and (b) agree.

Solution

(a) The allowed energy levels of the Bohr atom are given by:

$$\left(E_n\right) = -\frac{hcR}{n^2} = -\frac{me^4}{2\left(4\pi\varepsilon_0\right)^2\hbar^2 n^2} = -\frac{K}{n^2}$$

where K is given by K = hcR.

We designate the frequency of radiation emitted in a transition from state "n+1" to state "n" by $f(n+1 \rightarrow n)$. From the energy difference between the two levels we have:

$$hf(n+1 \rightarrow n) = E_{n+1} - E_n = -\frac{K}{(n+1)^2} + \frac{K}{(n)^2} = \frac{K(2n+1)}{(n)^2(n+1)^2}$$

(b) The radius of the nth Bohr orbit (r_n) (where a_0 is the radius of the first Bohr orbit) and the respective velocity (v_n) are:

$$r_n = n^2 a_0 \quad \text{and} \quad v_n = \frac{e^2}{2\varepsilon_0 hn}$$

Using the familiar relationship between linear velocity and angular velocity, namely $v = r\omega$; we obtain for the allowed angular velocity in a Bohr orbit:

$$\omega_n = (2\pi f_n) = \frac{v_n}{r_n} = \frac{e^2}{2\varepsilon_0 ha_0 n^3}$$

The frequency f_n is equal to:

$$f_n = \frac{e^2}{4\pi\varepsilon_0 ha_0 n^3} = \frac{K}{hn^3}$$

(c) We will express the frequency emitted in the transition $(n+1 \rightarrow n)$ in terms of the rotation frequency calculated in part (b); they are simply related:

$$f(n+1 \rightarrow n) = \frac{f_n n^3}{2} \frac{(2n+1)}{n^2(n+1)^2}$$

In the limit that $n \gg 1$, we can approximate $2n + 1$ by $2n$ and $n+1$ by n to obtain:

$$\frac{f(n+1 \rightarrow n)}{f_n} = \frac{n^3}{2} \frac{(2n+1)}{n^2(n+1)^2} \cong \frac{n^3}{2} \frac{2n}{n^2 n^2} = 1.$$

Example 9

An x-ray of energy 50 keV strikes an electron initially at rest. The x-ray is scattered through an angle of 90°. Calculate (a) the change in wavelength of the x-ray; (b) the energy of the x-ray after scattering; and (c) the velocity of the electron after scattering.

Solution:

Use the expression from the text relating the original wavelength λ to the wavelength after scattering λ' and the angle of scattering:

$$\lambda' - \lambda = \frac{h}{mc}(1 - \cos\theta)$$

For 90°, the wavelength change becomes

$$\lambda' - \lambda = \frac{h}{mc}$$

$$\lambda' - \lambda = \frac{(6.626 \times 10^{-34} \text{ J·s})}{(9.109 \times 10^{-31} \text{ kg})(3 \times 10^8 \text{ m/s})}$$

$$\lambda' - \lambda = 2.42 \times 10^{-12} \text{ m.}$$

The original wavelength can be calculated from the energy

$$hf = (5 \times 10^4 \text{ V})(1.6 \times 10^{-19} \text{ C}) = \frac{hc}{\lambda}$$

$$\lambda = \frac{c}{f} = \frac{(3 \times 10^8 \text{ m/s})}{(1.21 \times 10^{19} \text{ s}^{-1})} = 2.48 \times 10^{-11} \text{ m}$$

Solving for λ' results in

$$\lambda' = 2.48 \times 10^{-11} \text{ m} + 2.42 \times 10^{-12} \text{ m}$$

$$= 2.73 \times 10^{-11} \text{ m}$$

(b) The new frequency f' is calculated from

$$f' = (c/\lambda') = 1.10 \times 10^{19} \text{ Hz}$$

The x-ray energy after scattering is

$$hf' = (6.626 \times 10^{-34} \text{ J·s})(1.10 \times 10^{19} \text{ s}^{-1})$$

$$= 7.29 \times 10^{-15} \text{ J}$$

$$= 45.6 \text{ keV}$$

(c) The loss in x-ray energy appears as kinetic energy for the electron. Since the kinetic energy increase is much smaller than the electron rest energy, we can use the classical expression for the kinetic energy

$$\frac{1}{2} mv^2 = 50 \text{ keV} - 45.6 \text{ keV} = 4.44 \text{ keV}$$

To find the velocity, we can multiply and divide by c^2 to obtain

$$mc^2 \left(\frac{v}{c}\right)^2 = 2(4.44 \text{ keV})$$

Using $mc^2 = 0.511 \text{ MeV} = 511 \text{ keV}$ we find

$$\left(\frac{v}{c}\right)^2 = \frac{2(4.44 \text{ keV})}{511 \text{ keV}} = 1.74 \times 10^{-2}$$

$$v = (0.132)c = 3.96 \times 10^7 \text{ m/s}$$

Example 10

The explanation of the Compton effect for x-rays just makes use of general properties of photons and conservation laws. Why is this effect easily observable for x-rays but not for visible light?

Solution:

The effect would be observable for visible light but it is a small effect for light whereas it is quite dramatic for x-rays. The Compton formula is:

$$\lambda' - \lambda = \frac{h}{mc}(1 - \cos\theta)$$

We can rewrite this in terms of the frequencies f and f' as:

$$\frac{1}{f'} - \frac{1}{f} = \frac{h}{mc^2}(1 - \cos\theta)$$

or

$$\frac{f - f'}{f'} = \frac{hf}{mc^2}(1 - \cos\theta)$$

The angular factor $1 - \cos\theta$ can be at most 2, corresponding to the photon being "back-scattered". The photon energy is hf and the rest energy of the electron is mc^2 (about 0.5 MeV). For 550 nm light, the energy of the photon is about 2.26 eV. Thus $\Delta f = f - f'$ is:

$$\frac{\Delta f}{f'} \leq 10^{-5}$$

For x-rays, the frequencies are between 500 and 500,000 times larger than this "center frequency" for visible light, so the effect is much larger.

Example 11

Given the Planck expression for the spectral emittance:

$$I(\lambda) = \frac{2\pi hc^2}{\lambda^5 \left(e^{hc/\lambda kT} - 1\right)}$$

calculate the wavelength (λ_m) that maximizes $I(\lambda)$ and show that $\lambda_m T$ = constant.

Solution:

To maximize $I(\lambda)$, we will calculate its derivative and set that equal to zero and solve for the wavelength. It is actually easier to convert $I(\lambda)$ into a function of the frequency $f = c/\lambda$. Replacing λ by f gives:

$$I(f) = \frac{2\pi hf^5}{c^3 \left(e^{hf/kT} - 1\right)}$$

Taking the derivative with respect to frequency,

$$\frac{d}{df} I(f) = \frac{2\pi h}{c^3} \left[\frac{5f^4}{\left(e^{hf/kT} - 1\right)} - \frac{hf^5 e^{hf/kT}}{kT\left(e^{hf/kT} - 1\right)^2} \right] = 0$$

Canceling common terms and re-arranging leaves

$$5e^{hf/kT} - 5 = \frac{hf}{kT} e^{hf/kT}$$

Multiplying by $e^{-hf/kT}$ and defining $x = hf/kT$ reduces the equation to:

$$5 - x = 5e^{-x} \quad \text{or} \quad 1 - 0.2 x = e^{-x},$$

a transcendental equation that must be solved numerically. The right hand side, e^{-x}, as a function of x has its maximum value (unity) at $x = 0$ but it is always positive for finite x. The left hand side changes sign at $x = 5$ so the solution lies between $x = 0$ and $x = 5$.

The graph (Figure 38-2) where $f(x) = 1 - 0.2 x$ and $f(x) = \exp(- x)$ helps to locate the solution. The value of x where the two functions intersect is the solution.

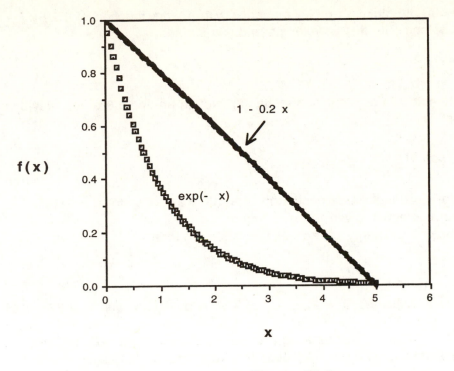

Figure 38-2

Since the solution lies very close to $x = 5$, to obtain sufficient precision, we can define Δ by $x = 5 - \Delta$ (where Δ *is small*) so that the equation to be solved becomes:

$$\Delta = 5e^{-5}e^{+\Delta}.$$

Since Δ is small, we can approximate the exponential $e^{+\Delta} \cong 1 + \Delta$, leading to a simple linear equation in Δ,

$$\Delta = (3.369 \times 10^{-2})(1 + \Delta).$$

Thus $\Delta = 3.486 \times 10^{-2}$ and $x = 5 - 3.486 \times 10^{-2} = 4.965$.

The frequency that produces a maximum in $I(\lambda)$ is then given by:

$$\frac{hf}{kT} = 4.965 = \frac{hc}{kT\lambda_m}$$

or $\lambda_m T = \text{constant} = hc/4.965k$.

QUIZ

1. Light of wavelength 500 nm strikes a metallic surface with a work function of 1.29 eV. Calculate the stopping potential for this wavelength and this surface.

Answer: $V_0 = 1.19$ volts

2. Given that the wavelength λ' after Compton scattering through an angle θ of an x-ray by an electron initially at rest is

$$\lambda' - \lambda = \frac{h}{mc}(1 - \cos \theta)$$

(a) Calculate the wavelength after scattering of a 100 KeV x-ray for $\theta = 180°$.
(b) Calculate the momentum given to the electron.
(c) Calculate the kinetic energy of the electron after scattering.

Answer: (a) $\lambda' = 1.73 \times 10^{-11}$ m,
            (b) 9.17×10^{-23} kg·m/s,
            (c) non-relativistic result $= 2.88 \times 10^4$ eV, relativistic result $= 2.82 \times 10^4$ eV

3. Light of wavelength $\lambda = 247.8$ nm strikes a pure cesium surface. The maximum kinetic energy of the ejected electrons is 3.20 eV. Calculate the work function of cesium.

Answer: The work function is 1.8 eV.

4. The electron in a hydrogen atom drops from an energy level where n = 4 to a level where n = 2. Calculate the energy and momentum of the emitted radiation.

Answer: The energy is 2.55 eV and the momentum is 1.36×10^{-27} kg·m/s

5. Suppose the charge on the nucleus in the Bohr atom is + Ze rather than + e, where Z is called the atomic number. Calculate the new energy levels E_n of the one electron Bohr atom.

Answer:

$$E_n = -\frac{Z^2}{n^2}(13.6 \text{ eV})$$

39
QUANTUM PHYSICS II: THE WAVE NATURE OF PARTICLES

OBJECTIVES

The objectives of this chapter are to:

Associate wave-like properties with material particles like electrons.

Calculate the wavelength associated with a material particle.

Apply the Bragg condition to electrons to understand electron diffraction.

Apply the uncertainty relationships to a variety of problems.

Define the wave function of a particle and interpret its absolute squared value as a relative probability.

REVIEW

The deBroglie hypothesis, applied to an atom, suggests that the allowed orbits should be those containing an integral number of electron wavelengths and gives identically the same results as the original Bohr model. However this was an important step forward in understanding fundamental processes. The association of a wavelength, λ, with a material particle (like an electron), where $\lambda = h/p$, was confirmed in the electron diffraction experiments of Davisson and Germer.

Bohr's postulates were used to obtain the energy levels of a one electron atom. The critical postulate is that the angular momentum must be an integral multiple of ($\hbar = h/2\pi$). Incorporation of that one idea into an otherwise classical framework produced a model of the atom with quantized *negative* energy levels (the zero of potential energy was chosen to coincide with infinite separation between the two charges) which forms the basis of understanding for the observed sharp line spectrum of atoms. In Example 2, the angular momentum quantization rule is applied to more familiar sized objects and it is shown there that it makes no difference to these ordinary motions.

The optical microscope has been duplicated using electrons rather than photons (light) leading to a standard research tool -- the electron microscope. In Example 3 these ideas are applied to a particle more massive than the electron, the neutron. Example 4 deals with electron diffraction.

One of the important features of the new theory called "Quantum Mechanics" is that the deterministic features of Newtonian mechanics are replaced by a statistical interpretation and there exist fundamental uncertainties in basic physical quantities like momentum, position, and time. The Heisenberg uncertainty relationships between linear momentum and position are:

$$\Delta p_x \Delta x \geq \frac{h}{2\pi} = \hbar \; ; \quad \Delta p_y \Delta y \geq \frac{h}{2\pi} = \hbar \; ; \quad \Delta p_z \Delta z \geq \frac{h}{2\pi} = \hbar$$

and the relationship between uncertainty in energy and uncertainty in time;

$$\Delta E \Delta t \geq \frac{h}{2\pi} = \hbar$$

These equations are usually stated as inequalities (rather than equalities) and interpreted in the following way: the product of the uncertainty in the x component of the linear momentum, Δp_x, and the uncertainty in the x coordinate, Δx, must be greater than or equal to Planck's constant (divided by 2π). This implies that a successively more accurate measurement of a particle's position (reduced Δx) introduces larger and larger uncertainty into the particle's linear momentum. These ideas are extended in Examples 8 and 9.

The uncertainty relationship connecting energy and time is illustrated in the text only to estimate the width, in energy, of a particular energy level if the average lifetime of that level is known. This is illustrated in Example 10.

In Newtonian mechanics, the state of a particle at any time could be specified by giving its velocity and its position at some initial time E. Newton's second law told us how this classical state changes in time. In quantum mechanics, all the information we can have about a particle is contained in its wave function. The absolute square (in many cases the wavefunction is a complex number) of the wavefunction evaluated at a particular point in space gives only the relative probability that the particle will be found at that point. The Schrodinger equation enables you to calculate the wavefunction for a given problem and also enables you to calculate the time dependence of this function.

QUESTIONS AND ANSWERS

Question. If tennis balls have wavelike properties, it might make for an interesting game, especially if there was destructive interference just before the served ball reached your opponent. Why doesn't this happen?

Answer. The wavelength of a tennis ball is too small (see Example 1) so that interference effects, which occur on the wavelength scale, are invisible.

Question. You are stopped for speeding and explain to the officer that by the uncertainty principle, he had no way of being certain just how fast you were going. Would this plea stand up before the jury of quantum mechanics?

Answer. No. Your speed uncertainty is $\Delta v = \hbar/m\Delta x$, where m is your mass and Δx is your position uncertainty. For any reasonable macroscopic measurement of Δx (such as 1 cm) and a mass in kilograms, the speed uncertainty is small compared to your speed because $\hbar$ is so small. Pay your fine or go to jail.

EXAMPLES AND SOLUTIONS

Example 1

Suppose an uncharged golf ball (m = 0.1 kg) teed from a hill went into orbit just at the (uncharged) earth's surface. Calculate the deBroglie wavelength associated with this material particle.

Solution:

Since the objects are uncharged, we will use the gravitational force between them as the dominant force. Then if the mass of the ball is m and M is the earth's mass,

$$\frac{GMm}{r^2} = \frac{mv^2}{r}$$

then $v^2 = GM/r = g\,(R_E)^2/r$ since $g\,(R_E)^2 = GM$. Since the ball is in orbit just at the earth's surface, $r \cong R_E$ so that $v^2 = gR_E$. This relationship enables us to calculate the linear momentum and then the deBroglie wavelength.

$$p^2 = (mv)^2 = m^2v^2 = m^2gR_E$$

Numerically we have;

$$p^2 = (0.1 \text{ kg})^2(9.8 \text{ m/s}^2)(6.4 \times 10^6 \text{ m}) = 6.27 \times 10^5 \,(\text{kg·m/s})^2$$

Therefore p = 792 kg·m/s. The wavelength is calculated from the value of p:

$$\lambda = \frac{h}{p} = \frac{(6.63 \times 10^{-34} \text{ J·s})}{(792 \text{ kg·m/s})} = 8.37 \times 10^{-37} \text{ m}$$

This wavelength is orders of magnitude smaller than the nucleus of the atom so that there is no known substance or material that could produce diffraction effects with such an object.

Example 2

Suppose the same uncharged golf ball (m = 0.1 kg) teed from a hill and placed into orbit just at the (uncharged) earth's surface is viewed as a "quantum object" in an orbit like the electron orbit in an atom. If its angular momentum is quantized as in the Bohr atom, what is the associated quantum number and what is the spacing between this allowed orbit and the next allowed orbit?

Solution:

Quantizing the angular momentum results in

$$mvr = n\hbar$$

so that $v^2 = (n\hbar/mr)^2$. In Example 1, we obtained the result that $v^2 = gR_E$ from classical considerations of the earth orbit. Equating the two equivalent expressions for v^2 and then solving for the allowed values of r, r_n, we have:

$$r_n = \frac{n^2h^2}{\left(2\pi mR_E\right)^2 g}$$

To estimate a value for the principle quantum number, n, set $r_n = R_E$ and solve for n,

$$n = \sqrt{\frac{g(2\pi m)^2 R_E^3}{h^2}}$$

Numerically we have

$$g = 9.8 \text{ m/s}^2, \qquad m = 0.1 \text{ kg,}$$

$$R_E = 6.4 \times 10^6 \text{ m and } \hbar = 1.05 \times 10^{-34} \text{ J·s;}$$

Thus

$$n \cong 5 \times 10^{43} \text{ !}$$

Since n is so large, the energy levels and allowed values of r are essentially continuous so to find the next value of r_n, say r_n+1, we write:

$$r_{n+1} = r_n + \left(\frac{dr}{dn}\right)\Delta n$$

Here $\Delta n = 1$ and the difference between r_{n+1} and r_n is the very small number,

$$r_{n+1} - r_n = \frac{2nh^2}{g\left(2\pi mR_E\right)^2}$$

Notice that with $r_n = R_E$ that:

$$\frac{r_{n+1} - r_n}{R_E} = \frac{2}{n} = 4 \times 10^{-44}$$

Since n is enormous, this fractional difference is totally negligible. It's no surprise that angular momentum quantization was not observed in terrestrial motions.

Example 3

What is the wavelength of a "thermal neutron" and what size of aperture would be needed to produce diffraction effects? For a "thermal neutron" the mean kinetic energy is $(3/2)k_BT$ where k_B is Boltzmann's constant and T is room temperature or about 300 K.

Solution:

Since the kinetic energy is equal to $p^2/2m$ and p is related to the wavelength by $p = h/\lambda$, we have:

$$\frac{1}{2m}\left(\frac{h}{\lambda}\right)^2 = \frac{3}{2}k_BT$$

Solving this for the wavelength, λ, we have

$$\lambda = \frac{h}{\sqrt{(3mk_BT)}}$$

Numerically the wavelength is

$$\lambda = \frac{\left(6.63 \times 10^{-34} \text{ J·s}\right)}{\sqrt{3(1.67 \times 10^{-27} \text{ kg})(1.38 \times 10^{-23} \text{ J/K})(300 \text{ K})]}}$$

$$= 0.145 \text{ nm}$$

Example 4

An electron accelerated through a potential difference of 1000 volts passes through a slit of width d, and then strikes a photographic film 0.5 m away. What size should d be so that the first minimum in the electron diffraction pattern on the screen will occur 0.1 mm from the center of the pattern?

Solution:

(a) To calculate the diffraction effects, we must know the electron's wavelength. From the potential difference, we can calculate the linear momentum, p, by:

$$eV = \frac{p^2}{2m}$$

$$p = \sqrt{2meV}$$

But $p = h/\lambda$ so the wavelength λ is obtained from:

$$\lambda = \frac{h}{\sqrt{2meV}}$$

with the result:

$$\lambda = \frac{\left(6.63 \times 10^{-34} \text{ J·s}\right)}{\sqrt{[2(9.1 \times 10^{-31} \text{ kg})(1.6 \times 10^{-16} \text{ J})]}}.$$

$$= 3.88 \times 10^{-11} \text{ m}$$

The first minimum in the diffraction pattern for a given wavelength, λ, is obtained when

$$\sin \theta = \lambda/d$$

where d is the aperture width. Using the small angle approximation for θ (valid since the vertical displacement, y, on the film is much smaller than the distance (D) from the aperture), the angle is $\sin \theta \cong \theta \cong y/D$, yielding

$$\frac{y}{D} = \frac{\lambda}{d}$$

or

$$d = \lambda \frac{D}{y}$$

The width is thus

$$d = \frac{\left(3.88 \times 10^{-11} \text{ m}\right)(0.4 \text{ m})}{\left(10^{-4} \text{ m}\right)} = 1.94 \times 10^{-7} \text{ m} = 194 \text{ nm}.$$

This is a slit width comparable to the wavelength of visible light and only several hundred times larger than the spacing between planes of molecules in a solid lattice.

Example 5

Taking the radius of the first Bohr orbit to be $r_0 = 0.053$ nm, (a) calculate the speed an electron would require for a wavelength equal to $2\pi r_0$ (the circumference of the orbit); and (b) the speed of an electron in the first Bohr orbit for an H atom. Compare the two answers.

Solution

(a) Using the de Broglie relationship $\lambda = h/p$, we equate $\lambda = 2\pi r_0$ and solve for v:

$$v = \frac{h}{m\lambda} = \frac{h}{m(2\pi r_0)} = \frac{\left(6.63 \times 10^{-34} \text{ Js}\right)}{\left(9.1 \times 10^{-31} \text{ kg}\right) 2\pi \left(5.3 \times 10^{-11} \text{ m}\right)} = 2.19 \times 10^6 \text{ m/s}$$

(b) The expression for v_n from the Bohr model where $n = 1$ is:

$$v_n = \frac{e^2}{2\varepsilon_0 hn}; \qquad v_1 = \frac{e^2}{2\varepsilon_0 h}$$

Numerically,

$$v_1 = \frac{e^2}{2\varepsilon_0 h} = \frac{2\pi e^2}{(4\pi\varepsilon_0)h} = \frac{2\pi \left(9 \times 10^9\right)\left(1.6 \times 10^{-19} \text{ C}\right)}{\left(6.63 \times 10^{-34} \text{ Js}\right)} = 2.19 \times 10^6 \text{ m/s}.$$

The two speeds are the same! (But we could have seen that analytically.)

Example 6

An electron with energy of 100 eV in region A of Figure 39-1, where the potential energy is zero, passes over a potential barrier in region B and a potential well in region C. Calculate the electron wavelength in regions A, B, and C.

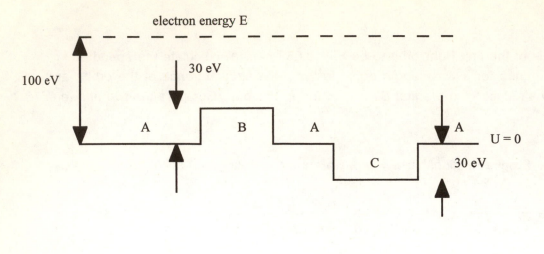

electron energy E

100 eV

30 eV

A   B   A   A   U = 0

C   30 eV

Figure 39-1

Solution

In region A we use the fact that the potential energy U = 0 to solve for the linear momentum

$$p_A = \sqrt{2m(E-U)} = \sqrt{2(9.1 \times 10^{-31} \text{ kg})(100 \text{ eV})(1.6 \times 10^{-19} \text{ J/eV})}$$

Numerically, $p_A = 5.4 \times 10^{-24}$ kgm/s. Next we use the de Broglie relationship $\lambda_A = h/p_A$ to find the associated wavelength:

$$\lambda_A = \frac{h}{p_A} = \frac{(6.63 \times 10^{-34} \text{ Js})}{(5.4 \times 10^{-24} \text{ kgm/s})} = 1.23 \times 10^{-10} \text{ m.}$$

In region B, the kinetic energy is reduced from its value of 100 eV in region A because the potential energy U = + 30 eV so E − U = 70 eV.

$$p_B = \sqrt{2m(E-U)} = \sqrt{2m(70 \text{ eV})} = p_A\sqrt{\frac{70 \text{ eV}}{100 \text{ eV}}}$$

$$\lambda_B = \lambda_A\sqrt{\frac{100}{70}} = 1.47 \times 10^{-10} \text{ m.}$$

In region A, again, the kinetic energy equals the total energy and the wavelength reverts to its original value. In region C, the potential energy is negative (U = − 30 eV) so the kinetic energy is now K = E − U = 130 eV. The new wavelength is:

$$\lambda_C = \lambda_A\sqrt{\frac{100}{130}} = 1.08 \times 10^{-10} \text{ m.}$$

Example 7

Suppose an electron is confined in an infinitely deep, one-dimensional, potential well of dimension L. Calculate the value of L required to make the photon frequency emitted in a transition from the n = 2 state to the n = 1 state equal to the photon frequency emitted in a transition from the n = 2 state to the n = 1 state in the Bohr hydrogen atom.

Solution

In the Bohr hydrogen atom model, the energy difference for the n = 2 to n = 1 transition is:

$$\Delta E_{H\;atom} = hf = E_2 - E_1 = -\frac{Rch}{(2)^2} - \left[-\frac{Rch}{(1)^2}\right] = \frac{3Rch}{4}$$

Since Rch = 13.6 eV, the above energy difference is ΔE = 10.2 eV.

For the particle-in-a-box energy levels for the infinitely deep well of length L,

$$E_n = \frac{n^2 h^2}{8mL^2}$$

The energy for the required transition is:

$$\Delta E_{e-box} = \frac{h^2}{8mL^2}\left(2^2 - 1^2\right) = \frac{3h^2}{8mL^2}$$

Equating this energy difference to that for the H atom and solving for L gives:

$$L = \sqrt{\frac{3h^2}{8m\Delta E_{e-box}}} = \sqrt{\frac{3h^2}{8m\left(10.2\;eV\right)}}$$

Solving numerically for L leads to:

$$L = \sqrt{\frac{3\left(6.63 \times 10^{-34}\right)^2}{8\left(9.1 \times 10^{-31}\right)\left(10.2\;eV\right)\left(1.6 \times 10^{-19}\;J/eV\right)}} = 3.3 \times 10^{-10}\;m.$$

If we compare this to the radius of the first Bohr orbit (r_0), we find that $L \cong 2\pi r_0$. Is that a surprise?

Example 8

A particle of mass m is to be confined by a potential well of depth V_0 and width d as shown in Fig. 39-2. For a given value of d what is the minimum value of V_0 necessary to produce confinement?

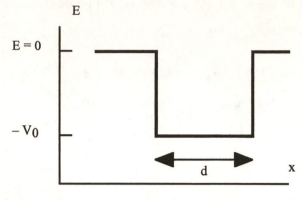

E = 0

$-V_0$

d

x

Figure 39-2

Solution:

To confine the particle to this well in one dimension requires that we know its position on the x axis with an uncertainty Δx equal to d. Using the relationship $\Delta x \, \Delta p = \hbar$, the uncertainty introduced into the x momentum is then:

$$\Delta p_x = \frac{\hbar}{\Delta x} = \frac{\hbar}{d}$$

This causes an uncertainty in the kinetic energy, ΔE_k, equal to:

$$\Delta E_k = \frac{(\Delta p_x)^2}{2m} = \frac{\hbar^2}{2md^2}$$

The total energy, E, of the particle must be less than zero for confinement (otherwise the particle could be anywhere on the x axis).

This condition can be stated as:

$$E = \Delta E_k - V_0 < 0$$

since the potential energy is equal to $-V_0$ for this problem. Solving this equation:

$$\frac{\hbar^2}{2md^2} < V_0 \quad \text{or that} \quad V_0 > \frac{\hbar^2}{2md^2}$$

308

Thus in one dimension as soon as a region of confinement (d) is specified, a potential energy depth V_0 can be calculated that will produce at least one bound state.

Example 9

What value of V_0 would be needed to produce a bound state in a nucleus of width $d = 3 \times 10^{-15}$ m for a proton or neutron (assume their mass to be the same)? Would this value of V_0 confine electrons inside the same well?

Solution:

We will use the inequality from the last example,

$$V_0 > \frac{\hbar^2}{2md^2}$$

For d we take the value given above and for m use 1.67×10^{-27} kg (for a proton or neutron). Then

$$V_0 \geq 3.67 \times 10^{-13} \text{ J} = 2.3 \text{ MeV}$$

Typical nuclear potential energies are about a factor of two lower than the above estimate but the "confinement" length increases slowly with nuclear mass number.

Since the electron mass is about 1836 times smaller than a proton or neutron mass, the value of V_0 needed to confine an electron in the nucleus would be increased by this factor of nearly two thousand. *It is safe to conclude that no electrons are confined within the nucleus.*

Example 10

The full width at half maximum intensity for a spectral line characteristic of a pH_2 molecule in an excited rotational energy level is 6×10^9 Hz. What estimate can be made of the lifetime of the molecule in this unstable state?

Solution:

To make this estimate, use the uncertainty relationship

$$(\Delta E)(\Delta t) \geq \hbar$$

The value of ΔE to be used is obtained from the frequency spread of the line Δf. Thus

$$\Delta E = h \Delta f$$

Substituting this into the uncertainty relationship yields

$$h(\Delta f)(\Delta t) \geq \hbar$$

or

$$\Delta t \geq \frac{1}{2\pi(\Delta f)} = \frac{1}{2\pi(6 \times 10^9 \text{ s}^{-1})}$$

$$\Delta t \geq 2.65 \times 10^{-11} \text{ s}.$$

QUIZ

1. An electron is accelerated from rest through a potential difference of 12 keV. Calculate the momentum after this acceleration and the associated wavelength.

Answer: (a) $p = 5.92 \times 10^{-23}$ kg·m/s; and (b) $\lambda = 1.12 \times 10^{-11}$ m.

2. A particle is confined by a "square well" potential of depth - V_0 and width $d = 2 \times 10^{-15}$ m.
    (a) How large must V_0 be in order to confine a proton ?
    (b) How large must V_0 be in order to confine an electron ?

Answer: (a) 5.2 Mev
            (b) 9.5×10^9 eV = 9.5 GeV.

3. A beam of electrons with kinetic energy of 100 eV strikes a pair of long narrow slits. The slits are separated by 10^{-6} m. Find the spacing between successive maxima in the interference pattern obtained on a screen 2 m from the slit system.

Answer: y = 0.245 mm

4. Assume that initially an electron is localized in a one dimensional region, $\Delta x_0 = 3 \times 10^{-10}$ m. Using the relationship $\Delta x\, \Delta p = \hbar$
(a) Calculate the spread in linear momentum for this electron.
(b) Calculate the corresponding spread in velocity.
(c) From part (b), calculate the spread in position after 0.2 s.

Answer: (a) $\Delta p = 3.5 \times 10^{-25}$ kg·m/s
           (b) $\Delta v = 3.86 \times 10^5$ m/s
           (c) $\Delta x = (\Delta v)t = 77$ km.

40
QUANTUM PHYSICS III: QUANTUM MECHANICS

OBJECTIVES

In this chapter, quantum mechanics is introduced and applied to simple but very general and useful systems. Your objectives are to:

Find wavefunctions, energy levels, and probability distributions for *a particle in a box* by fitting standing waves inside the box.

Construct the *Schrodinger equation* to produce the same levels and wave functions for a particle in a box.

Use the Schrodinger equation to study particles in *potential wells*, distinguishing between *bound states* and *free states*.

Find the *excitation energy* absorbed and emitted when a particle in a box or potential makes a transition from one energy level to another.

Calculate the *tunneling probabilities* for transmission through barriers.

Find the energies and wavefunctions of quantum *oscillators* and relate them to the energies of absorbed and emitted photons.

REVIEW

A "particle in a box" is one which cannot escape no matter how energetic it is and this is so in classical as well as quantum physics. In both cases the particle "sees" infinitely high potential "walls". The particle's probability must be zero there. When the walls are at $x = 0$ and L, these conditions require

$$\Psi_n = \sqrt{\frac{2}{L}} \sin \frac{n\pi x}{L} \quad \text{and } E_n = \frac{n^2 h^2}{8 \, mL^2} \quad \text{where } n = 1, 2, 3,...$$

Note that $\Psi_n(0) = \Psi_n(L) = 0$.

These are solutions of the Schrodinger equation

$$-\frac{\hbar^2}{2m}\frac{d^2\Psi}{dx^2} + U\Psi = E\Psi$$

when U = 0 (inside the box) subject to the boundary conditions $\Psi(0) = \Psi(L)$. The absolute square of the wavefunction $|\Psi(x)|^2$ is the probability density for finding the particle at x. The solutions Ψ_n have multiplicative constants arranged ("normalized") so that the probability of finding the particle anywhere between 0 and L is

$$\int_0^L |\Psi(x)|^2 \, dx = 1$$

In a finite potential well of depth U_0, (walls of height U_0) the wavefunction extends outside of the walls and the number of bound states ($E < U_0$) is finite and may be as few as one. They are also lower in energy than the corresponding level of a box of the same width.

In a finite well, particles are free and may have any energy greater than the well depth $E > U_0$. Energies are quantized only for $E < U_0$.

A potential barrier is an inverted well that repels particles incident upon it. The wavefunction of a particle is non-zero inside the barrier even if its energy is less than the barrier height. In classical physics, such a particle could not surmount the barrier and would be reflected back but in quantum physics the particle has a probability of penetrating or "tunneling" through the classically forbidden region. When the transmission coefficient or penetration probability T is small, it is approximately given by

$$T \approx e^{-2KL} \qquad \text{where } K = \frac{\sqrt{2m(U_0 - E)}}{\hbar}$$

where L is the barrier width and $\hbar K$ is the additional momentum the particle would need to be kicked to the top of the barrier.

The Schrodinger equation with an oscillator potential

$$U = \frac{1}{2} kx^2$$

may be solved to give the energy levels and wavefunctions. If the classical frequency of oscillation is:

$$\omega = \sqrt{\frac{k}{m}}$$

then the allowed quantum levels are:

$$E_n = \hbar\omega\left(n + \frac{1}{2}\right) \qquad \text{where } n = 0,1,2,...$$

and are equally spaced by an interval $\hbar\omega$ starting from the ground state (n = 0) at $E_0 = \hbar\omega/2$. many quantum systems (atoms in crystals, molecules) behave like oscillators when displaced slightly from their equilibrium points.

QUESTIONS AND ANSWERS

Question. Consider the allowed energy states of a "particle in a box". For any of these, compare the probability that the particle is to the left of center (of the box) to the probability that it is right of center (of the box).

Answer. These two probabilities are always the same because the probabilities are symmetric functions about the center point of the box.

Question. For any of the allowed energy states in Fig. 40-1, where are the places that the particle is more (least) likely to be found?

Answer. Most likely at L/6, L/2, 5L/6. Less likely at L/3, 2L/3.

EXAMPLES AND SOLUTIONS

Example 1

A particle in a box of width L is in the second excited state. (a) What is the probability of finding it between x = 0 and x = L/3? (b) What is the probability that it is within L/6 of one or the other of the walls?

Solution:

The energy levels have a ground state (n = 1) and first (n = 2) and second excited states (n = 3), etc. The wavefunctions are:

$$\Psi_n = \sqrt{\frac{2}{L}} \sin \frac{n\pi x}{L}$$

so that

$$\Psi_3 = \sqrt{\frac{2}{L}} \sin \frac{3\pi x}{L} .$$

This function has zeros at x = 0, x = L/3, x = 2L/3, and x = L. The probability density

$$|\Psi_3|^2 = \frac{2}{L} \sin^2 \frac{3\pi x}{L}$$

has three symmetric humps as shown in Fig. 40-1.

The total area under $|\Psi|^2$ yields the total probability

313

$$\int_0^L |\Psi_3|^2 \, dx = 1$$

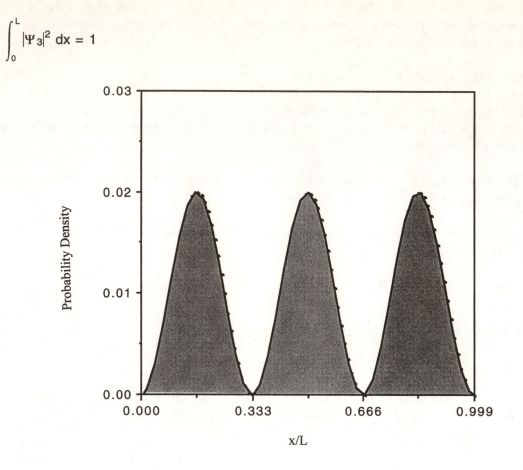

Figure 40-1

so that under each of the symmetric humps the areas are equal

$$\int_0^{L/3} |\Psi_3|^2 \, dx = \int_{L/3}^{2L/3} |\Psi_3|^2 \, dx = \int_{2L/3}^{L} |\Psi_3|^2 \, dx = \frac{1}{3}$$

(a) The probability of finding the particle in the first hump is 1/3 or 33%.

(b) The probability of finding the particle in the first half of the first hump or the second half of the third hump is:

$$P = \frac{1}{2}\frac{1}{3} + \frac{1}{2}\frac{1}{3} = \frac{1}{3} \quad \text{or } 33\%.$$

314

Example 2

An electron trapped in a box has a ground state energy of 10 eV. How big is the box?

Solution:

The ground state is the lowest energy state of the box,

$$E_1 = \frac{h^2}{8 \, mL^2} = 10 \text{ eV} = (10 \text{ eV})(1.6 \times 10^{-19} \text{ J/eV}) = 1.6 \times 10^{-18} \text{ J}.$$

Solving for the length L of the box,

$$L^2 = \frac{h^2}{8 \, mE_1} = \frac{\left(6.63 \times 10^{-34} \text{ J·s}\right)^2}{8 \left(9.11 \times 10^{-31} \text{ kg}\right)\left(1.6 \times 10^{-18} \text{ J}\right)}$$

or L = 1.94 x 10-10 m.

Example 3

Find the wavefunctions of a particle in a box centered at x = 0 with walls at x = ± L/2.

Solution:

The Schrodinger equation (for U = 0)

$$- \frac{\hbar^2}{2m} \frac{d^2\Psi}{dx^2} = E\Psi$$

has solutions

$$\Psi^{I} = \sin kx$$

$$\Psi^{II} = \cos kx$$

provided

$$E = \frac{\hbar^2 k^2}{2m}$$

which may be checked by direct substitution. Next we must satisfy the boundary conditions:

$$\Psi^{I}(L/2) = \Psi^{I}(-L/2) = 0$$

315

The boundary conditions limit the "allowed values" of k to the set

$$\frac{kL}{2} = m\pi \qquad m = 1,2, \ldots$$

$$k = \frac{2m\pi}{L} = \frac{n\pi}{L} \qquad n = \text{even}$$

The boundary conditions applied to the other wavefunction give

$$\Psi^{II}(L/2) = \Psi^{II}(-L/2) = 0$$

$$\cos\frac{kL}{2} = \cos\frac{-kL}{2} = 0$$

and limit the allowed values of k to the different set

$$\frac{kL}{2} = \frac{n\pi}{2}$$

$$k = \frac{n\pi}{L} \qquad n = \text{odd}$$

Thus the energy levels are

$$E_n = \frac{\hbar^2 k^2}{2m} = \frac{(h/2\pi)^2}{2m}\frac{n\pi^2}{L} = \frac{h^2 n^2}{8mL^2}$$

and the wavefunctions are

$$\Psi_n = \sin\frac{n\pi}{L}x \qquad n \text{ even; } n = 2,4,.$$

$$\Psi_n = \cos\frac{n\pi}{L}x \qquad n \text{ odd; } n = 1,3,\ldots$$

The first few are shown in Fig. 40-2.

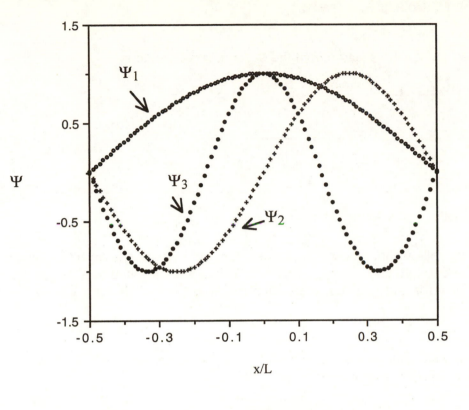

Figure 40-2

The first three wavefunctions have been plotted (without their normalization factor). Note they have the same shape as those for a box between x = 0 and x = L except they are displaced by L/2.

Example 4

An electron is trapped in a well of depth $U_0 = 6E_\infty$ and width L where E_∞ is the ground state energy of a particle in an infinitely deep well of the same width (particle in a box). Referring to Fig. 42-8 in the main text, what is (a) the ground state energy; and (b) the energy needed to excite the electron from the ground state to the first excited state; and (c) the energy needed to eject the electron from the well if it is in the second excited state.

Solution:

(a) The ground state energy for a well of this depth

$$E_1 = 0.625\ E_\infty = 0.625\ \frac{h^2}{8\ mL^2}$$

317

(b) The energy needed to excite the system from the ground state to the first excited state is:

$$E_2 - E_1 = 2.43 \ E_\infty - 0.625 \ E_\infty = 1.81 \ E_\infty = 1.81 \ \frac{h^2}{8 \ mL^2}$$

(c) The energy needed to free the electron from the second excited state is

$$\Delta E = U_0 - E_3 = 6 \ E_\infty - E_3$$

$$\Delta E = 6 \ E_\infty - 5.09 \ E_\infty = 0.91 \ E_\infty = 0.91 \ \frac{h^2}{8 \ mL^2}$$

Example 5

A car of mass 3000 kg rolls without friction on a level track and approaches a hill of height 1 m and width 1 m. It has enough kinetic energy so that classically it will climb to a height 0.5 m and then fall back. What is the quantum probability that it will tunnel through the hill?

Solution:

Since you have never witnessed such an event, the probability is certainly quite low so that the transmission coefficient can be approximated by

$$T \approx e^{-2KL}$$

where L = 1 m and K is given by

$$K = \frac{\sqrt{2m \ (U_0 - E)}}{\hbar}$$

In this case, the barrier height is

$$U_0 = mgY$$

and the energy is

$$E = \frac{mgY}{2}$$

so that

$$K = \frac{\sqrt{2m \ (mgY/2)}}{\hbar} = \frac{m\sqrt{gY}}{\hbar}$$

or solving for the product 2 KL,

318

$$2KL = \frac{2m\sqrt{g\gamma}\,L}{\hbar} = \frac{2(3000\ kg)\sqrt{(9.8\ m/s^2)(1\ m)}\,(1\ m)}{(1.05\times 10^{-34}\ J\cdot s)}$$

$$2KL = 1.8\times 10^{38}$$

Since T depends exponentially on this quantity, it is extremely small;

$$T \approx e^{-2KL} = \frac{1}{e^{1.8\times 10^{38}}} \ll 1$$

This probability is so small that the event is not worth waiting for.

Example 6

An atom is in the ground state of an oscillator potential. Estimate the typical deviation of an atom from its equilibrium position given its mass m and the natural frequency ω. Compare your result with the ground state wavefunction.

Solution:

In the ground state,

$$E_0 = \frac{\hbar\omega}{2}$$

For an oscillator the potential and kinetic energies typically (on average) are equal so that

$$E_0 \approx \overline{K} + \overline{U} \approx 2\overline{U} = 2\frac{1}{2}k(\overline{x})^2$$

where $\overline{x}$ is a typical displacement. Thus

$$E_0 = \frac{\hbar\omega}{2} = \frac{\hbar}{2}\sqrt{\frac{k}{m}} \approx k(\overline{x})^2$$

or

$$(\overline{x})^2 = \frac{\hbar}{2}\frac{1}{\sqrt{km}}$$

This may be compared with the ground state wavefunction

$$\Psi(x) = C\,e^{(-\sqrt{mk}\ x^2/2\hbar)}$$

$$\Psi(x) = C\,e^{(-x^2/4(\overline{x})^2)}$$

which is appreciable only for x comparable to or less than $\overline{x}$.

319

QUIZ

1. A proton is knocked out of a nucleus by a gamma ray (very energetic photon) of energy 6 MeV. Estimate the size of the nucleus if it is approximated by a box.

Answer: 5.85×10^{-15} m.

2. An electron has a 5% probability of tunneling through a barrier when its kinetic energy is 2 eV and the barrier height is 6 eV. What is the tunneling probability if the electron's energy is 4 eV?

Answer: 12%.

3. An atomic oscillator absorbs photons of energy about 0.1 eV when excited from the ground state to the first excited state. Estimate the typical vibrational distance in the ground state if the atom has a mass of 3.82×10^{-26} kg.

Answer: 10^{-10} to 10^{-9} m.

41

ATOMIC STRUCTURE

OBJECTIVES

The Schrodinger equation developed in last chapter is used to explain the experimentally observed electromagnetic radiation emitted by atoms. This explanation rests on the energy levels of the one-electron atom, the Pauli exclusion principle, and the existence of electron spin. In this chapter your objectives are to:

Develop the hydrogen atom wavefunctions and define the quantum numbers n, ℓ, and m_ℓ.

Apply the hydrogen atom wavefunctions to atoms containing many electrons.

Calculate the energy splittings known as the *Zeeman effect* from the magnetic interaction between a magnetic moment and an applied magnetic field.

Use the concept of electron spin to expand the number of quantum states for an atom for an atom described by the Schrodinger equation (n, ℓ, and m_ℓ) to four; namely (n, ℓ, m_ℓ, m_s).

Apply the Pauli exclusion principle to determine the electron configurations of the simpler atoms and the quantum numbers for atoms in their ground state.

Calculate the energy levels of multi-electron atoms using the central field approximation and the concept of electron screening.

Use experimentally known x-ray data to verify the quantum model of the atom.

REVIEW

This chapter applies the Schrodinger equation to calculate the allowed energy levels of the hydrogen atom. Restrictions placed on the solutions quantize the energy, the orbital angular momentum, and the component of angular momentum in the z direction. Each of these three quantization conditions produces one quantum number that can be associated with a specific energy state. To these three numbers, a fourth is added due to the existance of electron spin.

These discrete energy states in the one-electron atom, obtained from solutions of the Schrodinger equation, are used to indicate how more complicated atoms are formed. Their energies depend on nuclear charge (Z) resulting from the protons (each of charge + e) in the nucleus. Thus to obtain the correct expression for the energy, one replaces the factor e^2 by Ze^2

in the energy (or e^4 by Z^2e^4 etc.). Electron screening acts to reduce this effect.

THE HYDROGEN ATOM

The hydrogen atom wavefunctions and energies are used to explain the general features of atomic spectra. Those features were mainly found experimentally by observing photons emitted by atoms in the optical and x-ray regions of the electromagnetic spectra.

The hydrogen atom wavefunctions are factored into radial and angular components. The angular components contain the information about the quantized angular momentum while the radial component carries information about the energy and the electron probability of being observed in a particular region of space. See Example 1 where the electron probability distribution is investigated.

For the hydrogen atom, the energy levels are given by:

$$E_n = -\frac{(13.6\ \ eV)}{n^2}$$

where n is the *principal quantum number*, one of four quantum numbers to be introduced in this chapter to explain atomic spectra. *The energies do not depend on the other three quantum numbers if no external fields are present.*

Quantization of the (orbital) angular momentum results in the introduction of two more quantum numbers. The orbital angular momentum, L, is quantized in units of $\hbar$

$$L = \sqrt{\ell\left(\ell + 1\right)}\,\hbar$$

where ℓ, an integer, is restricted to be less than n the principal quantum number. In other words, $\ell = $ n-1, n-2,...0.

In addition, one other component, by convention the z component, is also quantized with

$$L_z = m_\ell \hbar \ \ \text{where } m_\ell = \ell, \ell -1,..., -\ell.$$

The z component cannot equal the total angular momentum due to the uncertainty principle. The general expression for the angle between these two quantities is given in Example 3.

Thus for a single value of n, there are many states, each state described by a unique combination of quantum numbers, that have the same energy. These states are said to be "degenerate". See Example 2 for a method of counting the number of degenerate states for a given n. *When the quantum numbers are large, quantum calculations agree with "classical" calculations.*

THE ZEEMAN EFFECT

When an atom is placed in a magnetic field, the degeneracy of the energy levels is partly removed. This is due to the fact that the orbital magnetic moment of the electron interacts with the magnetic field. The potential energy U is:

$$U = -\vec{\mu} \cdot \vec{B}$$

The fundamental unit of magnetic moment in atomic physics is the Bohr magneton, μ_B, where μ_B = 9.27 x 10^{-24} A/m^2 (or J/T). μ_B is not a new fundamental constant but is equal to $\mu_B = e\hbar/2m$, where e and m are the electron charge and mass respectively.

Taking B along the z axis results in splitting a single level into $2\ell + 1$ evenly spaced levels, one for each value of m_ℓ. This effect is called the *Zeeman effect*. The Zeeman effect can also be observed on a free electron, even one with no orbital angular momentum, since the electron has an intrinsic magnetic moment called "electron spin", but *only two levels are obtained.*

ELECTRON SPIN

The quantization of the angular momentum is a prediction of the Schrodinger equation. The quantum mechanical condition for quantizing the angular momentum permits an "angular momentum" to have values that are either integral or half-integral multiples of $\hbar = h/2\pi$. The usual kind of angular momentum, in the Bohr atom for instance, is called orbital angular momentum and has values that are only integral multiples of $\hbar$. A second kind of angular momentum exists called "spin angular momentum" of just *spin*.

Spin is a property of all the known particles with the electron, proton and neutron (for instance) having a spin of 1/2 while photons, deuterons and mesons have spins that are integral multiples of $\hbar$. The magnetic moment associated with electron spin is relatively large and leads to nearly all of the observed magnetic properties of bulk matter. Application of a magnetic field to a hydrogen atom causes one of the two spin 1/2 states to lie lower in energy than the other (because of the contribution to the energy of the term $-\vec{\mu} \cdot \vec{B}$) but in the absence of an external field these two levels have the same energy.

MANY ELECTRON ATOMS AND THE EXCLUSION PRINCIPAL

As more electrons are added to a complicated atom, they have a large effect on the atom's energy levels. The simplest scheme of incorporating the effect of the additional electrons on the energy consists of treating them like a spherically symmetric electron cloud surrounding the nucleus of charge + Ze so that the positive charge is partially neutralized. When this "central field" calculation is performed, the energy levels of the various atomic electrons depend not only on n, the principle quantum number, but on ℓ, the orbital angular momentum quantum number. If external electric and magnetic fields are present, the energies also depend on the orientation of both the orbital angular momentum and electron spin.

Several seemingly ad hoc rules are needed to explain atomic structure. First we need the Pauli exclusion principle which states for atoms that "no two electrons can occupy the same quantum state". Secondly we need to note that the specification of a "quantum state" actually depends on the problem under consideration but for an atom, four quantum numbers suffice to specify the state; n (the principle quantum number), ℓ (the orbital angular momentum quantum number), m_ℓ (the projection of the orbital angular momentum quantum number on the z axis), and m_s (the projection of the spin angular momentum on the z axis).

Thirdly, the quantum numbers n, ℓ, m_ℓ, and m_s have only certain ranges of values permitted. As previously stated, n is an integer 1, 2, 3 etc., and whatever value n takes, ℓ must be an integer less than n. For a given value of ℓ, m_ℓ can take on all values between $-\ell$ and $+\ell$ so there are $2\ell + 1$ possible values of m. For m_s the situation is simple; it can only be +1 or -1 and all it basically does to our discussion is to double the number of allowed states that would have been predicted for a given set of quantum numbers n, ℓ, m_ℓ, without any other effect. The quantum number m_s is very important for understanding the details of atomic structure and the magnetic properties of atoms, but we will not be concerned with the magnetic properties in this chapter. Finally, the atoms are formed by filling the electron levels with lowest energy first. For a given energy, the filling starts with s states and then *usually* proceeds with p, d, f, etc. in that order. These rules are illustrated in Tables 41-1 and 41-2. See Examples 1 and 2.

X-RAY SPECTRA

Photons of very high energy (in the keV range) have been detected in studies of atom emission spectra. These "x-rays" have guided the development of a correct model of the atom. They enable us to probe the energy level differences of the hydrogen atom levels in a multi-electron atom and verify the approximation used, called the central field approximation, in calculating their energies. See Examples 9 and 10.

QUESTIONS AND ANSWERS

Question. What problems are there in explaining atomic spectra and structure if the levels are not quantized?

Answer. Atoms would be unstable because, classically, electrons would radiate when in an orbit because of their accelerated motion. Atoms would also have *continuous spectra* as opposed to *discrete spectra* because any transition would be allowed.

Question. Consider a fictitious world in which there was no Pauli Principle. How would the atoms in that world differ from the atoms in our world?

Answer. The ground states of all atoms would be states with all the electrons in the same lowest energy state. There would not be a periodic table with the regularity we observe. Chemistry would be pretty boring.

EXAMPLES AND SOLUTIONS

Example 1

The probability of finding a particle within the radial range dr is given by:

$$P(r) \; dr = |\Psi|^2 \; dV = 4\pi r^2 \, |\Psi|^2 \; dr.$$

Find the value of r that maximizes P(r) for the hydrogen atom wavefunctions for the states n = 1, ℓ = 0 and n = 2, ℓ = 1.

Solution:

(a) For the state n = 1, ℓ = 0, we have the hydrogen-atom wavefunction:

$$\psi_{1s}(r) = \frac{1}{\sqrt{\pi a_0^3}} \, e^{-r/a_0}$$

The radial probability distribution, P(r), is then

$$P(r) = 4\pi r^2 \left| \frac{1}{\sqrt{\pi a_0^3}} \, e^{-r/a_0} \right|^2 = \frac{4\pi r^2}{\pi a_0^3} \, e^{-2r/a_0}$$

We take the derivative with respect to r and equate it to zero to locate any maxima or minima.

$$\frac{d}{dr} P(r) = \frac{4\pi}{\pi a_0^3} \left[2r \, e^{-2r/a_0} - \frac{2r^2}{a_0} e^{-2r/a_0} \right] = 0$$

This has the following factors:

$$(2r)\left(e^{-2r/a_0}\right)\left(1 - \frac{r}{a_0}\right) = 0$$

The possible roots are found by setting each of the factors equal to zero. Therefore the roots are: r = 0, r = ∞; and r = a_0. The roots r = 0 and r = ∞ correspond to minima but the root r = a_0 produces a maximum. *The radial probability distribution, for this wavefunction, peaks at the first Bohr orbit.*

(b) For the state n = 2, ℓ = 1, m_ℓ = 1, the hydrogen atom wavefunction is:

$$\Psi_{2p,1,1}(r) = - \frac{1}{\sqrt{8\pi a_0^3}} \left(\frac{r}{a_0}\right) e^{-r/2a_0} \sin\theta \; e^{i\phi}$$

To calculate P(r), we take the absolute square of this wavefunction, eliminating the dependence on angle ϕ. The proper element of volume is now dV = (r^2 dr)(sin θ dθ) dϕ. Integrating over ϕ

simply gives 2π. The term resulting in $\sin^2 \theta$ is then multiplied by $\sin \theta \, d\theta$ and integrated over θ leaving just those terms that depend on the radial coordinate. Collecting all the constants into the quantity K, we have for P(r),

$$P(r) = K \left(\frac{r}{a_0}\right)^4 e^{-r/a_0}$$

Using the same approach as before, we calculate the derivative:

$$\frac{d}{dr}P(r) = K\left[\frac{4}{a_0}\left(\frac{r}{a_0}\right)^3 e^{-r/a_0} - \frac{1}{a_0}\left(\frac{r}{a_0}\right)^4 e^{-r/a_0}\right] = 0.$$

This again has the roots $r = 0$, $r = \infty$ that produce minima but the new root that produces a maximum is $r = 4\, a_0 = (2)^2\, a_0$. It is generally true that the hydrogen atom wavefunctions where ℓ is the maximum possible value for a given n, correspond to the circular orbits of the Bohr atom and have radial distribution functions that peak at the values of the Bohr orbits.

The hydrogen atom wavefunction for $n = 2$, $\ell = 0$; the 2s state has a radial distribution function that does *not* peak at $2\, a_0$. Instead, it peaks at $r/a_0 = 3 \pm \sqrt{5}$!

Example 2

For a given hydrogen atom level specified by n, how many distinct states have the same energy? In other words, what is the degeneracy (g_n) of an atomic energy level of given principal quantum number n?

Solution:

In the Handbook of Chemistry and Physics, the sum of the first N integers is given by the expression:

$$S = 1 + 2 + 3 + \ldots\ldots + N = \sum_0^N n = \frac{N(N + 1)}{2}$$

For a given n, ℓ can take on all integral values from $\ell = n - 1$ to zero. Also, for a given ℓ, the magnetic quantum number m_ℓ can have $2\ell + 1$ values. Therefore we take all the values of m_ℓ and sum over the possible values of ℓ. Thus g_n is equal to:

$$g_n = \sum_{\ell=0}^{n-1} (2\ell + 1) = 2 \sum_{\ell=0}^{n-1} (\ell) + \sum_{\ell=0}^{n-1} (1) = 2\frac{(n - 1)n}{2} + n = n^2.$$

This counting does not include the degeneracy caused by electron spin. When that is taken into account g_n (as calculated above) is multiplied by a factor of two. Thus when all four quantum numbers are taken into account, in zero magnetic field, the degeneracy of an atomic energy level is $2\, n^2$.

Example 3

Develop the general formula for the angle between the total angular momentum and the z component and apply this formula to calculate the angle between them for n = 2,3,4.

Solution:

The maximum value of L_z is $\ell\hbar$, while the value of L is given by $\sqrt{\ell(\ell+1)}\cdot\hbar$. Therefore

$$\cos\theta = \frac{(L_z)_{max}}{L} = \frac{\ell\hbar}{\sqrt{\ell(\ell+1)}\,\hbar} = \frac{\ell}{\sqrt{\ell(\ell+1)}}$$

Numerically we can substitute the following values:

(1) When n = 2, $(L_z)_{max}$ = 1, $\cos\theta_1$ = 1/√(2); θ_1 = 45°.
(2) When n = 3, $(L_z)_{max}$ = 2, $\cos\theta_2$ = 2/√(6); θ_2 = 35.3°.
(3) When n = 4, $(L_z)_{max}$ = 3, $\cos\theta_3$ = 3/√(12); θ_3 = 30°.

Example 4

What simple physical considerations can be used to decide how the energy of an atomic orbital with given n and ℓ depends on the projection of ℓ, m_ℓ, and the spin projection, m_s, if there are no external fields applied?

Solution:

There appear to be none and for simple (low Z) atoms these levels all have about the same energy. In more complicated atoms, relativistic considerations (well beyond our scope) do make distinctions in these energy levels as do externally applied electric and magnetic fields.

Example 5

Write down the expected electron configuration for hydrogen atom states (a) for an atom with 18 electrons; (b) an atom with 22 electrons. (c) If an atom has 22 electrons, how many additional electrons would be required to complete a *closed shell*?

Solution:

(a) For an atom with 18 electrons, you start with the lowest energy state (1s) and place two electrons (electron spin up and electron spin down so the Pauli Principle is not violated) in that state leading to $1s^2$. Next you go to the next lowest energy state, 2s and place two electrons there leading to $1s^2 2s^2$. Now fill the remaining levels with n = 2, the 2p levels. There are three values of the magnetic quantum number (1,0, -1) since ℓ = 1, so there are *six* available states (again because of electron spin). Filling these states leads to $1s^2 2s^2 2p^6$, and 10 electrons have been used. This completes the second shell. For the third shell start with the 3s state and place two electrons there, place six electrons in the 3p orbitals and all 18 electrons have been used. The electron configuration for 18 electrons is then: $1s^2 2s^2 2p^6 3s^2 3p^6$.

(b) If there are 22 electrons available, then the first 18 are placed as in part (a) and the remaining four are placed in 3d orbitals where $\ell = 2$. Thus the electron configuration for 22 electrons is: $1s^2 2s^2 2p^6 3s^2 3p^6 3d^4$.

(c) Since $\ell = 2$ for the 3d states, there are $2(2\ell + 1) = 10$ total states in this level. From the electron configuration of 22 electrons, only 4 states were filled. This means that 6 more electrons are needed to fill this state and *complete the third shell*.

Example 6

The difference in energies of a hypothetical atom between its 2p and 3s levels is 1.2 eV. How large a magnetic field would be required to raise the energy of the highest possible state of the 2p level to that of the lowest possible 3 s state *due to the electron spin energies*?

Solution:

The energy difference is calculated from the potential energy of a magnetic moment in a magnetic field;

$$U = -\boldsymbol{\mu} \cdot \vec{\boldsymbol{B}}$$

where, for this example, we use the electron spin magnetic moment. The highest possible 2p energy due to the electron spin interaction with the magnetic field is:

$$E_{2p} = E_{2p}(0) + \mu_B B$$

where $E_{2p}(0)$ is the energy in the absence of the electron spin interaction with the field. The lowest possible 3s energy is:

$$E_{3s} = E_{3s}(0) - \mu_B B.$$

Therefore making $E_{2p} = E_{3s}$ for these states

$$E_{2p}(0) + \mu_B B = E_{3s}(0) - \mu_B B$$

or

$$E_{3s}(0) - E_{2p}(0) = 2\,\mu_B B = 1.2 \text{ eV}.$$

Solving numerically for the required magnetic field,

$$B = \frac{0.6 \text{ eV}}{\mu_B} = \frac{(0.6 \text{ eV})(1.6 \times 10^{-19} \text{ J/eV})}{(9.27 \times 10^{-24} \text{ J/T})} \approx 10^4 \text{ Tesla.}$$

This value is 500 times larger than a "high" steady magnetic field that can be produced with superconducting magnets. If we considered orbital angular momentum as well, this would reduce the field needed by 33% since the highest 2p level would have its energy increased by another $\mu_B B$.

Example 7

(a) Calculate the classical angular precession frequency of an electron in a constant magnetic field. (b) If the electron's angular momentum is quantized, calculate the allowed values of the possible orbits. (c) For an electron spin in a constant magnetic field, calculate the difference in energy of its two states and the corresponding angular frequency. (d) Compare this frequency with that calculated in part (a).

Solution:

(a) The classical angular frequency of precession for an electron (ω_c) is calculated from the forces; the magnetic force evB keeps the electron moving in a circular orbit of radius R;

$$F = evB = \frac{mv^2}{R} \quad \text{or} \quad \frac{eB}{m} = \frac{v}{R}$$

The classical frequency $\omega_c = v/R$ so

$$\omega_c = \frac{eB}{m}$$

(b) If the electron's angular momentum is quantized, $mvR = n\hbar$, where n is an integer. From the previous part (a), since $eBR = mv$, then $eBR^2 = mvR = n\hbar$. Therefore the allowed values of the radius are:

$$R_n^2 = \frac{n\hbar}{eB}$$

(c) For the electron spin potential energy, $U = -\vec{\mu} \cdot \vec{B}$ the energy difference ΔE will be:

$$\Delta E = 2\,\mu_B B = 2\,(9.27 \times 10^{-24} \text{ J/T})B.$$

In a magnetic field of B = 0.3 Tesla, the frequency ($hf = \Delta E$) needed to produce a transition between these two states is:

$$f = \frac{\Delta E}{h} = \frac{2\mu_B B}{h} = \frac{2(9.27 \times 10^{-24} \text{ J/T})(0.3 \text{ T})}{(6.63 \times 10^{-34} \text{ J·s})} = 8.39 \times 10^9 \text{ Hz} \cong 8.4 \text{ GHz.}$$

329

This is a commonly used frequency for electron paramagnetic resonance experiments now used extensively in chemistry and biology.

(d) The two frequencies are identical. This is seen by converting the classical angular frequency in $f_c = eB/2\rho m$ which equals the frequency calculated from the energy difference above.

Example 8

The x-ray transitions designated K_α and L_α are shown in Fig. 41-1. The energies of the x-ray photons, in keV, emitted in those transitions are tabulated below for five elements. Calculate from that data the energy differences between the levels $n = 2$ and $n = 1$ ($E_2 - E_1$) and $n = 3$ and $n = 2$ ($E_3 - E_2$).

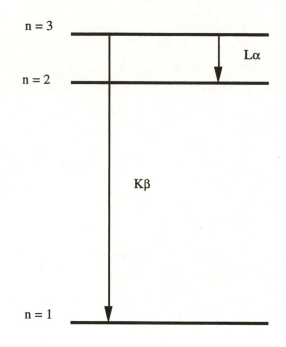

Figure 41-1

Solution:

The data is given in the following table where the positive charge on the nucleus of the element is denoted by Z:

Element	Z	K_α (keV)	L_α (keV)
Mn	25	6.51	0.721
Zn	30	9.57	1.11
Br	35	13.3	1.60
Zr	40	17.7	2.06
Rh	45	22.8	2.89

To find the energy differences (say $E_2 - E_1$) we use the definition of the x-ray lines. The K_α line arises from transitions from $n = 3$ to $n = 1$, thus it directly gives $E_3 - E_1$. The line L_α arises from transitions from $n = 3$ to $n = 2$, thus it gives directly the difference $E_3 - E_2$. The second part of the question is answered by listing the values of the L_α line, since $(E_3 - E_2) =$ energy of L_α Therefore the difference

$$E_2 - E_1 = (E_3 - E_1) - (E_3 - E_2) = \text{energy of } K_\alpha - \text{energy of } L_\alpha.$$

Mn, $E_2 - E_1 = 6.51$ keV - 0.721 keV = 5.79 keV.

Zn, $E_2 - E_1 = 9.57$ keV - 1.11 keV = 8.46 keV;

Br: $E_2 - E_1 = 13.3$ keV - 1.60 keV = 11.7 keV;

Zr: $E_2 - E_1 = 17.7$ keV - 2.06 keV = 22.8 keV;

Rh: $E_2 - E_1 = 22.8$ keV - 2.89 keV = 19.9 keV.

Example 9

Using the energy differences calculated in the previous example, calculate the effective Z value, Z_{eff}, from the central field approximation given by:

$$E_n = - \frac{Z_{eff}^2}{n^2}(13.6 \text{ eV})$$

for each of the five elements listed (a) using the energy differences $E_2 - E_1$; (b) using the energy differences $E_3 - E_2$; (c) using $Z_{eff} = Z - k$, for each series of energy differences, compute the average value of k. The quantity k measures the effectiveness of the screening of the inner electrons.

Solution:

(a) Using the above expression for the energy levels in the central field approximation, we can express the effective nuclear charge (the unscreened charge) in terms of the energy level differences found from the x-ray data. For instance:

$$E_2 - E_1 = - Z_{eff}^2 \left[\frac{1}{4} - \frac{1}{1} \right](13.6 \text{ eV}) = \frac{3}{4} Z_{eff}^2 (13.6 \text{ eV})$$

Solving for Z_{eff}, we have

$$Z_{eff}^2 = \frac{4}{3} \frac{(E_2 - E_1)}{(0.0136 \text{ keV})}$$

where the energy differences are in keV so 13.6 eV has been converted to keV. Using the values of $E_2 - E_1$ from the previous example gives:

Z_{eff} = 23.8, 28.8, 33.9, 39.2, and 44.2 respectively for Mn, Zn, Br, Zr, and Rh.

(b) For the second series generated by the energy differences $E_3 - E_2$; the expression for Z_{eff} becomes:

$$Z_{eff}^2 = \frac{36}{5} \frac{(E_3 - E_2)}{(0.0136 \text{ keV})}$$

leading to values of

Z_{eff} = 19.5, 24.2, 29.1, 33.0, and 39.1 respectively for Mn, Zn, Br, Zr, and Rh.

Note that these values are considerably smaller than the ones obtained from the energy differences between the lowest two levels.

(c) The values of $k = Z - Z_{eff}$ for the first series (denoted k_1) are:

$k_1 = 1.2, 1.2, 1.1, 0.8,$ and 0.8 respectively for Mn, Zn, Br, Zr, and Rh. Therefore the average value of $k_1 = 1.0$.

For the second series generated by the direct measured L_α lines, the values of $k = Z - Z_{eff}$ denoted k_2 are:

$k_2 = 5.5, 5.8, 5.9, 7.0,$ and 5.9 respectively for Mn, Zn, Br, Zr, and Rh. Therefore the average value of $k_2 = 6.0$.

Example 10

Suppose the electron has zero spin (it does not) so that a quantum state is specified by the three quantum numbers n, l, m only. If we also obey the rule that no two electrons can occupy the same quantum state (have the same quantum numbers), assign the electron configurations for the first ten elements in the periodic table. Identify the elements with "closed shell" configurations. {This would make for interesting chemistry.}

Solution

The assignments are given in the following table.

Element	Spin-zero Assignment	Correct Assignment	Comment on new chemistry
H (hydrogen)	$1s$	$1s$	Closed shell-inert
He (helium)	$1s2s$	$1s^2$	Very active
Li (lithium)	$1s2s2p$	$1s^2 2s$	
Be (beryllium)	$1s2s2p^2$	$1s^2 2s^2$	
B (boron)	$(1s2s2p^3)$	$1s^2 2s^2 2p$	Closed shell-inert
C (carbon)	$(1s2s2p^3)3s$	$1s^2 2s^2 2p^2$	Very active
N (nitrogen)	$(1s2s2p^3)3s3p$	$1s^2 2s^2 2p^3$	
O (oxygen)	$(1s2s2p^3)3s3p^2$	$1s^2 2s^2 2p^4$	
F (fluorine)	$(1s2s2p^3)3s3p^3$	$1s^2 2s^2 2p^5$	Closed shell-inert
Ne (neon)	$(1s2s2p^3)(3s3p^3)4s$	$1s^2 2s^2 2p^6$	Very active

QUIZ

1. Cu^{29} has one electron in a 4s state while Co^{27} has 2 electrons in a 4s state. How many 3d electrons are there in Cu^{29} and Co^{27}?

Answer: Cu has 10 (a complete shell) and Co has 7 (3 less than a complete shell).

2. Find the probability, by integrating the electron radial distribution function, of finding an electron in the 1s state with $r \geq a_0$.

Answer: $P = 5e^{-2} = 0.677$

3. In some semiconductors at very low temperatures, a one-electron atomic spectrum is observed. Assuming the effective charge on the nucleus is +e,
(a) Calculate the radius of the first Bohr orbit in a material where the relative dielectric constant $\varepsilon = 7$. (b) Calculate the frequency of radiation needed to remove an electron from the first Bohr orbit in the above material and drive it to infinity.

Answer: (a) $r(n = 1) = 3.71 \times 10^{-10}$ m, (b) $f = 6.73 \times 10^{13}$ Hz (in the infrared region).

4. An atom is in the state n = 5. What is the angle between the angular momentum L and the z component L_z?

Answer: $\theta = 26.6$ degrees.

5. What x-ray lines would be used to find the energy difference between the n = 4 and n = 2 levels?

Answer: The difference between K_α and L_α would provide the energy difference; or L_α would give it directly.

42
MOLECULES AND
CONDENSED MATTER

OBJECTIVES

In this chapter the formation of molecules from atoms and the formation of solids from molecules is described. Metals are discussed in terms of a *free-electron model* and the concept of band structure is combined with metallic behavior to describe two classes of conductors of high technological importance; semiconductors and superconductors. Your objectives are:

Classify the various molecular bonds as ionic, covalent, van der Waals, and hydrogen bonds.

Calculate the rotational energy levels of diatomic molecules and the frequencies emitted in transitions between rotational and vibrational levels.

Relate the electrical conductivity of a metal, insulator, or semiconductor to the molecular structure of its constituents and the resulting band structure.

Apply the free-electron model of a metal to pure metallic conductors, semiconductors, and even superconductors.

Compare the various types of solid order.

REVIEW

The formation of molecules from atoms can be qualitatively understood in terms of the energy levels in atoms calculated from quantum mechanics. The stability of closed shell configurations (since such a shell is of the inert gas form) makes it plausible that atoms with one electron outside such a shell would like to donate this electron to an atom lacking one electron to fill a shell. Such a charge exchange leads to the heteropolar bond -- an extremely favorable arrangement energetically. The binding energies associated with such ionic bonds are *much* larger than the van der Waals and hydrogen bonds and even larger than covalent bonds. The homopolar bond requires the sharing of electrons to create the bound state of the molecule. The Pauli exclusion principle leads to very symmetric electron cloud distributions in homopolar bonded molecules.

The spectrum of most molecules is confined to the region of the electromagnetic spectrum called the infrared. Interpretation of these spectra in terms of the spectra expected from a rigid rotating molecule or a molecule with vibrational modes like a simple harmonic oscillator (or a combination of the two) can provide valuable insight into the detailed structure of the molecule. See Examples 1, 2, and 3.

All of the naturally occurring elements except helium form solids at some temperatures without any externally applied pressure. In a crystalline solid, the molecules are organized into spatially repetitive patterns that are called lattices. The actual form of the lattice can be determined by either x-ray diffraction or neutron diffraction techniques as the spacing between planes in the lattice is comparable to the wavelengths of the projectiles used. Information about the collective state of molecules in a solid lattice can be obtained from measurements of the specific heat capacity, the thermal conductivity, the electrical conductivity, etc.

Based on the magnitude of the electrical conductivity, materials were historically classified as metals or insulators. Obviously semiconductors, as the name suggests, fall somewhere in between. Pure or intrinsic semiconductors have the same number of holes as electrons but in doped (impure) semiconductors the number of majority carriers per unit volume is much larger than the corresponding number of minority carriers. However the product of these two carrier densities in the doped semiconductor is equal to the corresponding product in the intrinsic semiconductor. The large disparity in carrier densities makes it relatively simple to construct a diode -- which is really just a one-way valve for electrical conduction. See Examples 4, 5, and 6.

The manner in which particles (such as electrons) with half-integral spin are distributed over the energy levels allowable to them is determined by the Fermi-Dirac distribution function (given in the main text). The valence electrons of elements that form metals are not tightly bound to their respective atoms and form a "gas" of mobile electrons or conduction electrons. This gas however obeys Fermi-Dirac statistics.

If, for instance, particle in a box energy states are used for the allowed levels, at the absolute zero of temperature, two electrons (one spin up and the other spin down) are placed in each state, starting from the lowest energy state, until all available electrons have been used. The highest energy of a filled state is called the Fermi energy (E_F).

The properties of the Fermi-Dirac distribution explain the general behavior of metals and form the basis of the "free electron theory of metals". Because the Fermi energy of most metals is of the order of a few electron volts (and one electron volt is the equivalent of 10,000 degrees Kelvin), only a small fraction of the electrons contribute to the thermal and paramagnetic properties of metals. See Examples 8 and 9.

Superconductors were discovered in the early part of the 20th century. In a superconductor, the electrical resistance drops from a finite value to zero at a finite temperature, T_c. The first element discovered to be a superconductor was mercury (Hg). Soon after that, it was found that lead and tin were also superconductors. Although the disappearance of the electrical resistance is a dramatic indication of a superconductor, the transition from the "normal" state to the "superconducting" state is a *magnetic phase transition*. In the superconducting state, magnetic lines of flux generated by an external field or source are excluded from the material by circulating surface currents. If the external field is large enough (equal to or larger than a

336

critical field B_c), the surface currents cannot null out the external field and the material reverts to the normal state. The critical field, (B_c) where the superconducting state is destroyed, is temperature dependent and modeled approximately by the expression:

$$B_c(T) = B_c(0)\left(1 - \frac{T}{T_c}\right)^2$$

where $B_c(0)$ is the critical field at 0 K (absolute zero) and T_c is the transition temperature in zero magnetic field.

Superconductors of the type first discovered are called "type I" and are the simplest to understand. It is characteristic of type 1 superconductors, that their critical temperature and their critical fields are relatively small. Another class of superconductors, with more complicated properties (many are alloys) have higher critical temperatures and higher critical fields. These are called "type II" superconductors. They are the constituents of most of the present day superconducting devices which require liquid helium for their operation. In 1987, superconductors with transition temperatures above 77 K, the boiling point of liquid nitrogen were discovered. Their properties are even more complicated and are the subject of active research at present.

Superconductivity is thought to result from a pairing of electrons; the formation of "Cooper pairs"-- electrons with equal and opposite momenta (hence same energy) and opposite spins-- not unlike the pairs of electrons that populate the Fermi levels of a metal. These pairs however have a *correlated* space motion with the distance (or correlation length) very large. See Examples 10 and 11.

QUESTIONS AND ANSWERS

Question. What kind of chemical bond holds together the following objects: (a) NaCl molecules, (b) N_2 molecules, and (c) Cu atoms in a wire.

Answer. (a) Ionic bonds, (b) covalent bonds, (c) metallic bonds.

Question. How does the effectiveness of a pn junction diode vary as the voltage difference across the junction gets smaller?

Answer. The forward and backward resistances become more equal and the diode becomes less effective.

EXAMPLES AND SOLUTIONS

Example 1

The homonuclear molecule D_2 is made up of two deuterons (a proton plus a neutron) and two electrons (as in H_2). The spacing between the nuclei in the molecule is approximately 7.5×10^{-11} m. Assuming the masses of the electrons are negligible compared to the nuclear masses, calculate the moment of inertia of this molecule about the center of mass and the rotational energy in electron volts of the ground state and the first two excited states.

Solution:

The moment of inertia is

$$I = \Sigma \, m_i r_i^2.$$

If we denote the mass of the proton by by m_H, the mass of the deuteron is about $2m_H$. If d is the separation, then each value of r is d/2 so that:

$$I = 2 \, m_H \left(\frac{d}{2}\right)^2 + 2 \, m_H \left(\frac{d}{2}\right)^2 = m_H d^2$$

Numerically since $m_H = 1.67 \times 10^{-27}$ kg and $d = 7.5 \times 10^{-11}$ m, we have,

$$I = 9.39 \times 10^{-48} \text{ kg·m}^2$$

The energy levels are given by:

$$E = (\hbar)^2 \frac{L(L+1)}{2 I} = (5.92 \times 10^{-22} \text{ J}) \, L(L+1)$$

$$= (3.70 \times 10^{-3} \text{ eV}) \, L(L+1)$$

The ground state (state with lowest energy is obviously the state with $L = 0$ so $E_0 = 0$. The first excited state has $L = 1$ so $E_1 = 7.40 \times 10^{-3}$ eV. The second excited state has $L = 2$ so $E_2 = 2.22 \times 10^{-2}$ eV.

Example 2

The hydrogen isotopes (H, D, and T) form three heteronuclear molecules, HD, HT, and DT. In each case, the internuclear separation is d = 7.5 x 10⁻¹¹ m. Take the mass of H (m_H)to be the proton mass m_p = 1.67 x 10⁻²⁷ kg and $m_D = 2m_H$, $m_T = 3 M_H$. (a) Calculate the moments of inertial of these three molecules about their center of mass. (b) Calculate for each molecule the ratio

$$T_R = \frac{\hbar^2}{kI}$$

where k is the Boltzmann constant and I is the moment of inertial from part (a). This quantity, with dimensions of a Kelvin temperature, characterizes the energy level separation of the respective molecules.

Solution:

For the homonuclear molecules, the center of mass is midway between the two nuclei but this is not so for HD, HT, and DT.

(a) The position of the center of mass is given by:

$$R_c = \frac{\Sigma \; m_i r_i}{\Sigma \; m_i}$$

HD: If we measure this distance from the H nucleus in HD, then R_c = 2d/3 and

$$I_c = m_p\left(\tfrac{2}{3}d\right)^2 + 2m_p\left(\tfrac{1}{3}d\right)^2 = \tfrac{2}{3}m_pd^2$$

HT: Again measuring from the H nucleus in HT, R_c = 3d/4 and

$$I_c = m_p\left(\tfrac{3}{4}d\right)^2 + 3m_p\left(\tfrac{1}{4}d\right)^2 = \tfrac{3}{4}m_pd^2$$

DT: Measuring from the D nucleus in DT, R_c = 3d/5 and

$$I_c = 2m_p\left(\tfrac{3}{5}d\right)^2 + 3m_p\left(\tfrac{2}{5}d\right)^2 = \tfrac{6}{5}m_pd^2$$

Note that each of these expressions is equivalent to $I_c = \mu d^2$ where μ is the "reduced mass" of the molecule.

(b) The "rotational temperature", T_R, is found to be:

$$T_R = \frac{\hbar^2}{kI} = \frac{3\hbar^2}{2km_pd^2} \qquad \text{for HD}$$

The factor $\hbar^2/km_p d^2$ will be common to all these molecules. Numerically it is equal to:

$$\frac{\hbar^2}{km_p d^2} = \frac{\left(1.05 \times 10^{-34} \text{ J}\cdot\text{s}\right)^2}{\left(1.38 \times 10^{-23} \text{ J/K}\right)\left(1.67 \times 10^{-27} \text{ kg}\right)\left(7.5 \times 10^{-11} \text{ m}\right)^2} = 85.9 \text{ K}$$

Therefore, for HD, $T_R = (3/2)(85.9 \text{ K}) = 129$ K. For HT, $T_R = (4/3)(85.9 \text{ K}) = 114$ K. For DT, $T_R = (5/6)(85.9 \text{ K}) = 71.6$ K. Note for H_2, $T_R = 172$ K and for D_2, $T_R = 85.9$ K.

Example 3

(a) If a molecule with moment of inertia I is induced to make a pure rotational transition from a state L to a state L + 1, what frequency of radiation is needed?

(b) If the same molecule makes a transition from the state L to the state L - 1, what frequency of radiation is emitted?

Solution:

(a) Since the energy is

$$E = \hbar^2 \frac{L(L + 1)}{2I}$$

in going from L to L + 1, we must supply energy, hf, equal to:

$$hf = \hbar^2 \frac{1}{2I}\left[(L + 1)(L + 2) - L(L + 1)\right]$$

$$hf = \hbar^2 \frac{2(L + 1)}{2I}$$

(b) If the molecule drops down from the level characterized by L to the level L - 1, the emitted radiation frequency, f', is given by:

$$hf' = \hbar^2 \frac{[L(L + 1) - (L - 1)L]}{2I}$$

$$hf' = \hbar^2 \frac{2L}{2I}$$

Note: the spacings between the possible values of f and f' will be the same.

340

Example 4

The current-voltage relationship for a p-n junction is given as:

$$I = I_0[\exp(eV/kT) - 1]$$

Obtain an expression for the ratio of the "forward" resistance to the "backward" resistance. Evaluate this ratio when the voltage is 0.2 volts and T is 293 K.

Solution:

The "forward" resistance is that resistance obtained when the battery polarity is such that the current is large.

$$R_F = \frac{V}{I} = \frac{V}{I_0\left[e^{eV/kT} - 1\right]}$$

The backward resistance, R_B, is obtained when the above voltage polarity is reversed. The current direction is also reversed so:

$$R_B = \frac{-V}{I} = \frac{-V}{I_0\left[e^{-eV/kT} - 1\right]}$$

$$\frac{R_F}{R_B} = -\frac{\left[e^{-eV/kT} - 1\right]}{\left[e^{+eV/kT} - 1\right]}$$

Numerically we have

$$\frac{eV}{kT} = \frac{(1.6 \times 10^{-19})(0.2)}{(1.38 \times 10^{-23})(293)} = 7.91$$

Since $\exp(7.91) = 2.74 \times 10^3$, the above ratio is approximately the inverse of this number or $R_F/R_B = 3.6 \times 10^{-4}$.

Example 5

Given that the current in a semiconductor p-n junction is

$$I = I_0\left(e^{V_b/kT} - 1\right)$$

(a) show that the current can be approximated by

$$I \approx \frac{I_0 V_b}{kT}$$

when V_b is small compared to kT. How small must V_b be for this approximation to be valid? (b) If $I_0 = 1.44 \times 10^{-9}$ A, calculate the "reverse bias resistance". (c) If the junction is forward biased by $V_b = +0.259$ volts, calculate the current and the resistance. (d) Evaluate the "dynamic resistance" (dV/dI) for the conditions of part (c).

Solution:

(a) If the quantity $V_b/kT \ll 1$, then the exponential can be expanded since $e^x = 1 + x + ...$,

$$I = I_0\left(1 + \frac{V_b}{kT} + ... - 1\right) \approx \frac{I_0 V_b}{kT}$$

At 300 K (near room temperature) the factor $kT = 2.59 \times 10^{-2}$ eV. For the expansion to be valid, $V_b \ll 0.026$ volts.

(b) In the very small reverse bias voltage; $R_{eff} = V_b/I$;

$$R_{eff} = \frac{V}{I} = \frac{kT}{I_0} = \frac{0.0259 \ \text{volts}}{1.44 \times 10^{-9} \ \text{amps}}$$

$$R_{eff} = 18 \ M\Omega$$

(c) If the junction is forward biased by 0.259 volts, $V_b/kT = 10$. Thus

$$I = (1.44 \ nA)\left(e^{10} - 1\right) \cong (1.44 \ nA)e^{10} = 3.17 \times 10^{-5} \ A.$$

The resistance in this case is:

$$R_f = \frac{V_b}{I} = \frac{0.259 \ V}{3.17 \times 10^{-5} \ A} = 8170 \ \Omega.$$

This is 2200 times smaller than the reverse bias resistance from part (b).

(d) The dynamic resistance is defined as

$$R_{dynamic} = \frac{dV}{dI} = \frac{kT}{I_0 e^{V_b/kT}} = \frac{0.0259 \ V}{\left(1.44 \times 10^{-9} \ A\right)\left(2.2 \times 10^4\right)} = 818 \ \Omega.$$

Example 6

Electromagnetic radiation can promote an electron from the top of a nearly complete valence band to the bottom of an unfilled conduction band. The lowest frequency for which this is possible is $f_m = 2.75 \times 10^{14}$ Hz for Si and $f_m = 1.79 \times 10^{14}$ Hz for Ge at room temperature. Calculate the energy gap bwteen the valence and conduction bands of Si and Ge based on this data.

Solution:

The minimum energy between the two bands is the energy gap E_{gap}. Converting the photon frequency into an energy,

$$\Delta E = hf_m = E_{gap}$$

Therefore for Si

$$E_{gap} = \frac{\left(6.63 \times 10^{-34} \ J \cdot s\right)\left(2.75 \times 10^{14} \ s^{-1}\right)}{\left(1.6 \times 10^{-19} \ J/eV\right)} = 1.13 \ eV$$

Correspondingly for Ge

$$E_{gap} = \frac{\left(6.63 \times 10^{-34} \ J \cdot s\right)\left(1.79 \times 10^{14} \ s^{-1}\right)}{\left(1.6 \times 10^{-19} \ J/eV\right)} = 0.736 \ eV$$

Example 7

A system of N_A particles, where N_A is Avogadro's number, each with electron spin of 1/2 is placed in a static magnetic field **B**. (a) Ignoring any interactions between the particles, calculate the average "spin" energy using Boltzmann statistics and approximate this result for the case where $\mu_B B \ll kT$. (b) Using the approximate spin energy, calculate the "spin" heat capacity.

Solution:

The potential energy for a magnetic moment μ in a magnetic field is $U = - \vec{\mu} \cdot \vec{B}$. For an electron spin this becomes:

$$U = - \vec{\mu} \cdot \vec{B} = - m_s \mu_B B$$

where $m_s = \pm 1$ and μ_B is the Bohr magneton. Thus there are two energy levels per spin; $E_+ = \mu_B B$ and $E_- = - \mu_B B$. The relative probabilities, using Boltzmann statistics, of finding the particle in these states are:

$$p(+) = \frac{e^{- \mu_B B/kT}}{e^{- \mu_B B/kT} + e^{+ \mu_B B/kT}} \quad \text{and} \quad p(-) = \frac{e^{+ \mu_B B/kT}}{e^{- \mu_B B/kT} + e^{+ \mu_B B/kT}}$$

(a) Therefore the average "spin" energy of a single particle, $<E>$, is:

$$<E> = (\mu_B B)p(+) + (-\mu_B B)p(-).$$

Since the particles are "non-interacting", the total energy is the sum of the individual particle energies or $E_{total} = N_A <E>$.

If $\mu_B B \ll kT$, then the exponential factors are all small and can be approximated by

$$e^x \cong 1 + x + \frac{x^2}{2} + \dots$$

where we will keep only the first two terms and neglect the quadratic and higher terms. Thus

$$p(+) \cong \frac{1}{2}\left(1 - \frac{\mu_B B}{kT}\right) \quad \text{and} \quad p(-) \cong \frac{1}{2}\left(1 + \frac{\mu_B B}{kT}\right)$$

and E_{total}, the "spin energy contribution" is approximately equal to:

$$E_{total} \cong (-1)(N_A \mu_B B)\left(\frac{\mu_B B}{kT}\right)$$

(b) The contribution to the heat capacity made by the spins is found by taking the derivative of E_{total} with respect to temperature (at constant magnetic field),

$$C_{spin} = \frac{d}{dT}E_{total} \cong (N_A \mu_B B)\left(\frac{\mu_B B}{kT^2}\right)$$

Example 8

Suppose for a particle-in-a-box, the energy values are given by:

$$E_n = E_0 n^2$$

where E_0 is a constant and n is the quantum number of the state. If there are 50 electrons in such states, (a) find the Fermi energy at a temperature of zero Kelvin. (b) What is the ratio of the average energy of the electrons to the Fermi energy?

Solution

(a) Putting two electrons per state according to the Pauli Principle, we occupy 25 states. Therefore the energy of the 25th state is the Fermi Energy at T = 0.

$$E_{F0} = E_0 (25)^2 = 625\, E_0$$

(b) The total energy is obtained by summing the contributions from each of the states.

$$E_T = 2E_0[1 + 4 + 9 +...] = 2E_0\left(\sum_1^{25} n^2\right)$$

The general sum over the first N squares of integers is:

$$\sum_1^N n^2 = \frac{N(N+1)(2N+1)}{6}$$

Evaluating this for N = 25 gives the total energy as:

$$E_T = 2E_0[1 + 4 + 9 +...] = 2E_0[(25)(13)(17)]$$

Since there are 50 particles, the average energy per particle is:

$$e_{avg} = \frac{E_T}{50} = (221)E_0 = \left(\frac{221}{625}\right)E_{F0}$$

Example 9

Given the Fermi-Dirac probability distribution, and the fact that Sodium has a fermi energy of 3.15 eV, find the ratio of the width ΔE to E_F at T = 273 K where $\Delta E = E(0.2) - E(0.8)$.

The quantity $E(0.2)$ is the energy where the occupation probability is 0.2.

Solution

Given that $f(E)$ is:

$$f(E) = \frac{1}{1 + e^{(E-E_F)/kT}}$$

then by solving for E we obtain:

$$E = E_F = kT \ln\left(\frac{1 - f(E)}{f(E)}\right)$$

Substituting for the $f(E)$ values,

$$E(0.2) = E_F + kT \ln 4; \quad \text{and} \quad E(0.8) = E_F - kT \ln 4$$

Therefore for the ratio to the fermi energy we obtain:

$$\frac{\Delta E}{E_F} = \frac{2kT \ln 4}{E_F} = \frac{2\left(8.617 \times 10^{-5} \text{ eV}/\text{K}\right)\left(273 \text{ K}\right) \ln 4}{\left(3.15 \text{ eV}\right)} \cong 0.02$$

This means that near room temperature, an energy spread (ΔE) equal to 2% of the Fermi energy changes the occupation probability for Sodium from 80% to 20%, a very sharp function.

Example 10

The element tin (Sn) has 1.48×10^{23} conduction electrons per cm^3. The atomic weight of tin is 118.7 and its density is 7.30 g/cm^3. (a) How many electrons does each atom contribute to the conduction band? (b) The critical magnetic field at $T = 0$ is 0.0305 Tesla. Calculate the energy difference *per pair of electrons* between the normal state and the superconducting state at $T = 0$. (c) The known energy gap in Sn, $E_{gap} = 1.15 \times 10^{-3}$ eV, is representative of the energy difference between the normal and superconducting states. Using the calculated energy difference between the two states from part (a) and the known energy gap for Sn, how many pairs of electrons must have their energy reduced by E_{gap} to account for the superconductivity? (d) The Fermi energy for tin is 10.2 eV. How many pairs of electrons are there with energies within kT (where for T the value of $T_c = 3.72$ K is used) of the Fermi energy? How does this number compare with the number of pairs used in part (b) and that calculated in part (c)?

Solution:

(a) Let N_A = the number of atoms (of Sn) per cm^3 and n = the number of conduction electrons per unit volume. The number of electrons per atom is then the ratio $n = N_e/N_A$. The number of atoms per unit volume is found from the data on tin given:

$$N_A = \left(6.02 \times 10^{23} \text{ atoms/mol}\right)\left(7.30 \text{ g/cm}^3\right)\left(\frac{1 \text{ mole}}{118.7 \text{ g}}\right) = 3.70 \times 10^{22} \text{ atoms/cm}^3$$

Therefore the number of electrons per atom is:

$$n = \frac{N_e}{N_A} = \frac{1.48 \times 10^{23} \text{ electrons/cm}^3}{3.70 \times 10^{22} \text{ atoms/cm}^3} \cong 4 \text{ electrons/atom}$$

(b) Just at the phase boundary between the normal and superconducting states, an applied magnetic field below the critical field cannot penetrate the superconductor but can the normal metal. Thus the energy difference between the two states per unit volume is given just by the magnetic energy per unit volume:

$$\Delta E = E_n - E_s = \frac{B_c{}^2}{2\mu_0} = \frac{(0.0305)^2}{2(4\pi \times 10^{-7})} = 370 \frac{J}{m^3} = 3.70 \times 10^{-4} \frac{J}{cm^3}$$

Since 1 eV = 1.6×10^{-19} J, $\Delta E = 2.31 \times 10^{15}$ eV/cm^3. From the value of N_e in part (a) we find that there are $N_e/2$ pairs. Therefore the energy difference per pair is:

$$\frac{\Delta E}{N_e/2} = \frac{(2.31 \times 10^{15} \text{ eV/cm}^3)}{(7.4 \times 10^{22} \text{ pairs/cm}^3)} = 3.12 \times 10^{-8} \text{ eV/pair.}$$

This "correlation energy" is so small that the superconducting state would be very fragile and easily destroyed by thermal motions.

347

(c) For this part, we equate the product of the number of pairs and the energy gap to the magnetic energy calculated in part (a).

$$N_{pairs}E_{gap} = \frac{B_c^2}{2\mu_0} = 2.31 \times 10^{15} \frac{eV}{cm^3}$$

Using the given value of E_{gap},

$$N_{pairs} = \frac{2.31 \times 10^{15} \ eV/cm^3}{1.15 \times 10^{-3} \ eV} = 2.01 \times 10^{18} \ per \ cm^3.$$

This number is 10^4 smaller than we used in part (c) when we calculated the number of pairs from the conduction electron density.

(d) Let N_s be the number of electrons within kT of the Fermi energy. The Fermi energy E_F can be expressed as a temperature T_F by writing:

$$T_F = \frac{E_F}{k_B} = \frac{(10.2 \ eV)(1.6 \times 10^{-19} \ J/eV)}{(1.38 \times 10^{-23} \ J/K)} = 1.18 \times 10^5 \ K.$$

Therefore we obtain for N_s (using T_c as the temperature):

$$N_s \cong N_e\frac{k_BT_c}{k_BT_F} = \frac{(1.48 \times 10^{23} \ electrons/cm^3)(3.72 \ K)}{(1.18 \times 10^5 \ K)} = 4.66 \times 10^{18} \ electrons/cm^3$$

Thus the number of pairs is: $N_{pairs} = 2.33 \times 10^{18}$ pairs/cm3. This is very close to the value calculated in part (c), suggesting that only those electrons near the Fermi energy are participants.

Example 11

The material Nb$_3$Sn (niobium-tin) is a "type II" superconductor frequently used in the construction of high magnetic field superconducting magnets. It has the following characteristics: the critical temperature Tc = 18.0 K; the critical field B$_c$(0) = 24.5 Tesla.
(a) Fing the current needed to produce a field on the axis of a *single layer* solenoid wound from wire of Nb$_3$Sn, with 1 mm diameter, of 10 Tesla. (b) Calculate the magnetic field at the surface of a wire of diameter 1 mm equal carrying the current calculated in part (a). to one-half the critical field at a temperature of 4.2 K (the boiling point of liquid helium). (b) Find the field on the axis of a *single layer* solenoid wound from wire of Nb$_3$Sn, with 1 mm diameter, carrying this current. (c) How does this calculated value compare with the critical field at 4.2 K?

Solution:

(a) The relationship between magnetic field and current for a solenoid is:

$$B = \mu_0 n I$$

where n is the number of turns per unit length and I is the current in each turn. Since the wire diameter is 1 mm, there can be 1000 turns per meter, so n = 1000. Substituting numerically:

$$10 \text{ T} = (4\pi \times 10^{-7})(1000)\text{ I}$$

or I = 7960 amperes.

(b) The field at the surface can be calculate from Ampere's law applied to a long straight wire:

$$B(r) \cdot 2\pi r = \mu_0 I$$

Using r = 5 x 10^{-4} m and I = 7960 amperes, yields B$_{surface}$ = 3.18 T.

(c) Using the relationship between critical field and temperature:

$$B_c(T) = B_c(0)\left(1 - \frac{T}{T_c}\right)^2$$

we substitute the data on Nb$_3$Sn to obtain:

$$B_c(4.2 \text{ K}) = (24.5 \text{ T})\left(1 - \frac{4.2 \text{ K}}{18.0 \text{ K}}\right)^2 = 14.4 \text{ Tesla.}$$

When the field at the wire surface is B$_{surface}$ = 3.18 T; this represents 22 % of the critical field at 4.2 K.

QUIZ

1. (a) Calculate the potential energy in eV of an H^- ion separated by 7.5×10^{-11} m from a H^+ ion. This would be the bond energy for an ionic bonded state of the H_2 molecule. (b) The bond energy of H_2 is -4.5 eV. Is the molecular bond of H_2 ionic or covalent?

Answer: (a) $U = -19.2$ eV; (b) The H_2 bond is covalent.

2. The separation between the nuclei H and D in the molecule HD is 0.074 nm (as in H_2 or D_2). Calculate (a) the moment of inertia of this molecule about its center of mass and (b) the frequency of radiation emitted in a pure rotational transition from the $L = 1$ state to $L = 0$ state.

Answer: (a) $I = 6.06 \times 10^{-48}$ kg·m², (b) $f = 2.77 \times 10^{12}$ Hz.

3. In some semiconductors at very low temperatures, a one-electron atomic spectrum is observed. Assuming the effective charge on the nucleus is $+e$,
(a) Calculate the radius of the first Bohr orbit in a material where the relative dielectric constant $\varepsilon = 7$. (b) Calculate the frequency of radiation needed to remove an electron from the first Bohr orbit in the above material and drive it to infinity.

Answer: (a) $r_1 = 3.71 \times 10^{-10}$ m, (b) $f = 6.73 \times 10^{13}$ Hz (in the infrared region).

4. An atom of hydrogen with kinetic energy of 10 eV collides with an H_2 molecule originally in its lowest vibrational level. The fundamental vibrational frequency of an H_2 molecule is 1.29×10^{14} Hz. If the atom transmits all or part of its kinetic energy to the vibrational energy of the molecule, find the possible vibrational states produced in the collision.

Answer: States up to and including $n = 18$ can be produced.

5. Show that for a two dimensional system of N particles, the quantity $g(E) = dN/dE$ is *independent of E* and calculate the resulting Fermi energy.

Answer: $E_F = (N/A)\pi\hbar^2/m$ where A is the area and m is the mass.

43

NUCLEAR PHYSICS

OBJECTIVES

In this chapter, aspects of nuclear physics are introduced. Your objectives are to:

Recognize the *nucleus* as a small core (size about 10^{-14} m) inside the larger atom (size about 10^{-10} m).

Apply *tables of nuclear data* and atomic data (Young and Freedman, Table 43-1 and 43-2) to practical problems. Distinguish between mass number (A), atomic number (Z) and neutron number (N) and use the notation $_ZX^A$ where X is chemical name for the nucleus.

Calculate *mass defect* and *binding energies* for the nuclei in these tables.

Compare actual binding energies with *liquid drop model* binding energies.

Calculate the kinetic *energy released* in α, β and γ *radioactive decays*.

Calculate *abundances* and *activities* for a radioactive source given the decay constant or the half-life.

Discuss the biological effects of radiation in terms of energy absorbed.

Find the kinetic energy which must be supplied to, or is released in, a *nuclear reaction*.

Distinguish between nuclear *fission* and nuclear *fusion*.

REVIEW

The nucleus is composed of Z protons and A-Z neutrons where A is the mass number and Z is the atomic number. The charge on the nucleus is + Ze. There are Z electrons surrounding the nucleus in a neutral atom.

Table 43-2 (text by Young and Freedman) lists the atomic mass, including the atomic electrons, in the unit u = 1.660566 x 10^{-27} kg. The nuclear mass is

$$m(_ZX^A) = \text{atomic mass} - Zm_e$$

where Z is the number of electrons.

The mass of the constituent nucleons (protons and neutrons) is

$$Zm_p + (A-Z)m_n$$

The mass defect Δ is

$$\Delta = Zm_p + (A-Z)m_n - m(_ZX^A)$$

and is also called the binding energy when expressed in energy units. The binding energy per nucleon is Δ/A in energy units.

In the liquid drop model of the nucleus the binding energy E_B is given by

$$E_B = C_1A - C_2A^{2/3} - C_3\frac{Z(Z-1)}{A^{1/3}} - C_4\frac{(A-2Z)^2}{A}$$

where C_1, C_2, C_3, and C_4 are constants.

Nuclei can decay in various ways, for example by falling apart into:

(1) another nucleus plus an α particle (α-decay)

(2) another nucleus plus an electron plus a neutrino (β-decay)

(3) another nucleus plus a photon (γ-decay)

The decay is possible only if the kinetic energy release is positive. The kinetic energy released is equal to the difference between the mass of the parent nucleus and the sum of the masses of the decay products. The neutrino is massless. The mass of the α particle, a helium nucleus $_2He^4$, and other nuclei may be calculated from Table 45-2 (text by Young) by subtracting the mass of the atomic electrons. The stability of a nucleus against α and β decay may be checked in this way to see if the released kinetic energy is positive; if it is not, kinetic energy must be supplied to make the reaction go, and the nucleus is stable against the given decay.

The number of unstable nuclei in a sample declines as

$$N = N_0\exp(-\lambda t)$$

where λ is the decay constant, related to the "half-life" by

$$T_{1/2} = \frac{\ln 2}{\lambda} = \frac{0.693}{\lambda}.$$

Since the activity A is the number of decays per unit time - dN/dt, we have also

$$A = A_0\exp(-\lambda t)$$

The Relative Biological Effectiveness (RBE) of radiation on biological systems depends on the dose (energy delivered per unit mass) multiplied by a factor that takes into account the type of radiation depositing the energy.

In nuclear *fission* a large nucleus is induced, usually by neutron bombardment, to fall apart into two smaller nuclei. If this reaction itself produces neutrons which in form trigger another fission reaction, a chain reaction is said to occur and the process is sometimes self-sustaining. This is the principle behind a nuclear reactor or a bomb. In both cases energy is released.

Another energy release reaction is the *fusion* reaction in which two light nuclei (e.g. hydrogen or helium) combine to form a heavier one, with the release of energy. An application of this reaction for power generation is being intensely studied but as yet no practical scheme has been found.

PROBLEM-SOLVING STRATEGIES

Please refer to the problem-solving strategy for Nuclear Properties in the main text. Many of the problems of this chapter involve energy unit conversion:

$$1 \text{ eV} = 10^{-6} \text{ MeV} = 1.6 \times 10^{-19} \text{ J}$$

$$1 \text{ u} = \text{atomic mass unit} = 931.5 \text{ MeV}/c^2 = 1.66 \times 10^{-27} \text{ kg}$$

A curie is a measure of the radioactive decay intensity.

$$1 \text{ curie} = 3.7 \times 10^{10} \text{ decays per second}$$

If a radioactive decay or reaction equation is given, Table 43-2 (text by Young) may be used to calculate the mass of each side of the equation, term by term. To balance the energy conservation equation, supply kinetic energy to the appropriate side of the equation. Remember to subtract the atomic electron masses from the atomic masses of Table 43-2 (text by Young) if the nuclear mass is desired.

Some problems involve photons or γ rays. Recall that the energy of a photon is

$$E = pc = \frac{hc}{\lambda}$$

A rad is a unit of absorbed energy: 1 rad = 0.01 J/kg.

QUESTIONS AND ANSWERS

Question. When light elements fuse together, energy is released. Is energy released when heavy elements combine?

Answer. No. Heavy elements undergo fission and release energy; to keep the energy balance, energy has to be added to fuse the nuclei together that form the heavy elements.

Question. Knowing that light elements release energy when they fuse and that heavy elements release energy when they undergo fission, compare the binding energy per nucleon for light, medium, and heavy nuclei.

Answer. The binding energy per nucleon is large for light nuclei, smaller for medium nuclei, and large for heavy nuclei.

EXAMPLES AND SOLUTIONS

Example 1

The density of nuclear matter is about 2×10^{17} kg/m³. Find the radius R of an iron nucleus (A = 56).

Solution:

The density is the mass per unit volume

$$\rho = \frac{M}{\frac{4}{3}\pi R^3}$$

where M = 56 u, if the binding energy is neglected. Thus the radius is given by

$$R = \left(\frac{M}{\frac{4}{3}\pi\rho}\right)^{\frac{1}{3}}$$

$$R = \left(\frac{56 \times 1.66 \times 10^{-27} \text{ kg}}{\frac{4}{3}\pi \cdot 2 \times 10^{17} \text{ kg/m}^3}\right)^{\frac{1}{3}} = 4.8 \times 10^{-15} \text{ m}.$$

354

Example 2

An alpha particle with 5 MeV kinetic energy makes a head on collision with a silver nucleus. Find the distance of closest approach between the alpha particle and nucleus.

Solution:

If the alpha particle and nucleus stay far enough apart so that nuclear forces are negligible, only the force of coulomb repulsion acts. The conserved energy is

$$E = K + U = \frac{1}{2} mv^2 + \frac{Q_1 Q_2}{4\pi\varepsilon_0 r} \text{ ,}$$

where r is the separation. The charges are given by the charge on the α particle, $Q_1 = 2e$ and the charge on the silver nucleus, $Q_2 = 47e$.

When the distant α particle begins its motion at $r = \infty$ the initial energy is

$$E_1 = \frac{1}{2} mv^2 = 5 \text{ MeV} = 5 \times 10^6 \text{ eV}$$

$$= (5 \times 10^6 \text{ eV})(1.6 \times 10^{-19} \text{ J/eV}) = 8 \times 10^{-13} \text{ J}$$

If the final energy is taken to be the point of closest approach, when the velocity is zero, we have

$$E_2 = \frac{Q_1 Q_2}{4\pi\varepsilon_0 r} = \frac{(2e)(47e)}{4\pi\varepsilon_0 r} = \frac{94 \ e^2}{4\pi\varepsilon_0 r}$$

Since the energy is conserved; meaning $E_1 = E_2$

$$8 \times 10^{-13} \text{ J} = \frac{94 \ e^2}{4\pi\varepsilon_0 r}$$

$$r = \frac{94 \ e^2}{4\pi\varepsilon_0 \left(8 \times 10^{-13} \text{ J}\right)}$$

$$r = \frac{94 \left(1.6 \times 10^{-19} \text{ C}\right)^2}{4\pi\left(8.9 \times 10^{-12} \text{ C}^2 \cdot \text{N}^{-1} \cdot \text{m}^{-2}\right)\left(8 \times 10^{-13} \text{ J}\right)}$$

$$= 2.7 \times 10^{-14} \text{ m}$$

This is close to the radius of the nucleus, where nuclear (non-coulomb) forces begin to act.

Example 3

Find the mass defect, the total binding energy, and the binding energy per nucleon for N^{14}.

Solution:

Referring to Table 43-2 in the text, the atomic mass of N^{14} is 14.00307 u. The mass of the N^{14} nucleus is obtained by subtracting from this the mass of the seven electrons (Z = 7) in the atom

$$m(N^{14}) = 14.00307 \text{ u} - 7(0.000549 \text{ u})$$

$$= 13.99923 \text{ u}$$

The mass of the seven protons (Z = 7) and seven neutrons (A-Z = 14 - 7 = 7) is

$$7 \, m_n + 7 \, m_p = 7(1.008665 + 1.007276) \text{ u}$$

$$= 14.11159 \text{ u}$$

The mass defect is thus

$$7 \, m_n + 7 \, m_p - m(N^{14}) = (14.11159 - 13.99923) \text{ u} = 0.124 \text{ u}$$

Since 1 u = 931.5 MeV the mass defect or total binding energy is

$$0.124(931.5 \text{ MeV}) = 104.7 \text{ MeV}$$

The binding energy per nucleon is

$$\frac{104.7 \text{ MeV}}{14} = 7.5 \text{ MeV per nucleon.}$$

Example 4

(a) Find the mass of the tritium nucleus $_1H^3$ if its atomic mass is 3.01647 u.
(b) Find the mass of the helium nucleus $_2He^3$ plus an electron at rest.
(c) If a tritium nucleus decays into a helium nucleus plus an electron plus a neutrino, what energy is released as kinetic energy?

Solution:

(a) The mass of the $_1H^3$ nucleus is the atomic mass from Table 43-2 (text by Young) less the mass of a single electron,

$$m(_1H^3) = 3.01647 \text{ u} - 0.000549 \text{ u}$$

$$= 3.01592 \text{ u}$$

(b) By Table 43-2 (text by Young), atomic helium has a mass of 3.01603 u. By subtracting the mass of the two atomic electrons we find the mass of the nucleus,

$$m(_2He^3) = 3.01603 \text{ u} - 2(0.000549 \text{ u})$$

$$= 3.014932 \text{ u}$$

The mass of this nucleus plus the mass of a single electron at rest is

$$m(_2He^3) + m_e = 3.014932 \text{ u} + 0.000549 \text{ u} = 3.015481 \text{ u}$$

(c) In the decay

$$_1H^3 = {_2He^3} + \text{electron} + \text{neutrino}$$

The kinetic energy of the electron and the neutrino is the left over mass or 'mass defect',

$$K(\text{electron} + \text{neutrino}) = m(_1H^3) - m(_2He^3) - m_e$$

$$= 3.01592 \text{ u} - 3.015481 \text{ u}$$

$$= 4.4 \times 10^{-4} \text{ u} = 0.41 \text{ MeV}$$

Example 5

The half-life of C14 is 5568 years. If C14 dating was attempted for a piece of wood believed to be 2000 years old, what is the abundance of C14 in the wood compared to wood of the same kind which has been freshly cut from a tree?

Solution:

The number of C14 nuclei, N, decreases with time according to

$$N = N_0 \exp(-\lambda t)$$

The half-life $T_{1/2}$ is related to λ by

$$\lambda = \frac{0.693}{T_{1/2}} = \frac{0.693}{5568 \ \text{yr}} = 1.24 \times 10^{-4} \ \text{yr}^{-1}$$

If N is the number of C14 (for equal mass) in the new wood, the old wood has the fractional abundance

$$\frac{N}{N_0} = \exp(-\lambda t) = \exp\left[-\left(1.24 \times 10^{-4} \ /yr\right)(2000 \ \text{yr})\right] = 0.78$$

Example 6

Co60 decays to Co59 with a half-life of 5.3 years.
    (a) What is the activity, in curies, of a source containing 0.015 g of Co60?
    (b) What is the activity of the source 2 years later?

Solution:

(a) The number of Co60 atoms initially present is

$$N_0 = \frac{(0.015 \ \text{g})}{(60 \ \text{g/mol})}\left(6.02 \times 10^{23} \ /\text{mol}\right) = 1.5 \times 10^{20}$$

The activity is the number of decays per unit time. At the initial time we have

$$\left|\frac{dN}{dt}\right|_{t=0} = \left|\frac{d}{dt}\left(N_0 \ e^{-\lambda t}\right)\right|_{t=0} = \left|\lambda N_0 \ e^{-\lambda t}\right|_{t=0} = \lambda N_0$$

$$\left|\frac{dN}{dt}\right|_{t=0} = \frac{0.693}{T_{1/2}} N_0 = \frac{0.693}{5.3 \ \text{yr}}\left(1.5 \times 10^{20}\right) = 1.96 \times 10^{19} \ /\text{yr}$$

$$\left|\frac{dN}{dt}\right|_{t=0} = \frac{1.96 \times 10^{19} \ /\text{yr}}{(365)(24)(3600) \quad \text{s}} = 6.2 \times 10^{11} \ /\text{s}$$

Since a curie is 3.7 x 10^{10} decays per second, the activity in this unit is

$$\lambda N_0 = \frac{6.2 \times 10^{11} \text{ /s}}{3.7 \times 10^{10} \text{ /s}} = 16.9 \text{ curies}$$

(b)   In two years the activity will be

$$\left|\frac{dN}{dt}\right| = (\lambda N_0) \, e^{-\lambda t} = (16.9 \text{ curie}) \exp\left[-(0.63)(2 \text{ yr})/5.3 \text{ yr}\right]$$

$$= 13 \text{ curies}$$

Example 7

A 100 kg worker receives an annual dose of x-ray radiation of 5 rem. (a) How much energy is deposited in her system? (b) If all this energy liberated neutrons from her nuclei, how many would be liberated? (c) How does this compare to her total number of nucleons?

Solution:

(a) For x-rays, RBE = 1, so

$$1 \text{ rem} = 1 \text{ rad} = 0.01 \text{ J/kg}$$

or a total of 1 J is deposited.

(b) The binding energy per nucleon is about 8 MeV/nucleon.

$$1 \text{ J} = \frac{10^{25}}{1.6} \text{ MeV}$$

So the energy could liberate

$$\frac{\left(\frac{10^{25}}{1.6} \text{ MeV}\right)}{(8 \text{ MeV})} = 7.8 \times 10^{23} \text{ nucleons}$$

(c) The worker has

$$\frac{100 \text{ kg}}{\left(1.66 \times 10^{-27} \text{ kg}\right)} = 6.0 \times 10^{28} \text{ nucleons}$$

so roughly 10 nucleons per million could be liberated.

359

QUIZ

1. Suppose that the proton decayed into a positron (particle with mass of the electron but opposite charge) and a photon. Find the kinetic energy released in MeV.

Answer: 938 MeV

2. For how many half-lives would one have to wait before the activity of a radioactive source declines to 1/100 of its original value?

Answer: 3.3

3. How much energy is required to release a neutron from the nucleus $_8O^{17}$, forming $_8O^{16}$. Refer to Table 45-2 of the text.

Answer: 4 MeV

4. The binding energy per nucleon for the elements peaks around mass number $A = 60$ where it is 8.7 MeV and drifts down to 7 MeV near $A = 250$. Suppose a nucleus of mass number 240 fragmented into 4 nuclei of mass number 60. Estimate the total energy that would be released.

Answer: 408 MeV

44
PARTICLE PHYSICS AND COSMOLOGY

OBJECTIVES

In this final chapter you are introduced to the elementary building blocks of matter, their forces and interactions, and their role in the early history of the universe. Your objectives are to:

Recognize protons and neutrons as the constituents of nuclei.

Recognize that the positron, like the electron in every respect but opposite in charge, is but one example of an *antiparticle*. All particles have antiparticles.

Associate forces among elementary particles with other particles that are emitted and absorbed by the original particles.

Find the radius of rotation and the angular frequency of charged particles in cyclotrons and synchrotrons.

Classify the four known forces of nature, and the particles that mediate them.

Distinguish between Leptons and Hadrons and conserve lepton number in reactions.

Build up the hadrons from constituent quarks with electric charges (e/3) and (2e/3) and their antiparticles.

Define "The Standard Model" of Particle Interactions.

Describe the "big bang theory" as a model for the expanding universe.

Describe the "Standard Model" of the history of the universe.

Explain how dark matter and dark energy affect the expansion rate of the universe.

REVIEW

This chapter covers the topics in a qualitative way. However many simple calculations based on what we have learned during the entire course shed light on issues as basic as the elementary nature of matter and the earliest moments of time. Many of these calculations depend only on the conservation of energy and momentum and the relation between matter and energy;

$$E = mc^2 = \sqrt{p^2c^2 + m_0^2c^4}$$

When equal amounts of matter m and antimatter m combine to produce pure energy--for instance when an electron and a positron combine at rest to produce a photon--the energy is $2\ mc^2$.

At one time protons, neutrons, electrons, and neutrinos were thought to belong to a select group of elementary and indivisible particles. Further study indicated that protons and neutrons are but two members of a large class of particles (hadrons, including many other baryons and mesons) that are themselves composed of more basic strongly interacting particles called quarks. Electrons and neutrinos are still elementary and indivisible as far as can be experimentally determined, but also are but two members of a large class of particles called leptons. Hadrons undergo strong, electromagnetic, and weak interactions. Leptons do not have strong interactions. All particles are subject to gravitation.

Elementary particle interactions conserve electric charge and energy and also other "charge-like" quantum numbers like *baryon number* and *lepton number*. Baryons are composed of three quarks and an antiquark.

QUESTIONS AND ANSWERS

Question. A hydrogen atom is neutral like a neutron and has about the same mass. In what ways are the two systems very different?

Answer. The hydrogen atom has nearby excited electronic energy levels that are a tiny fraction of its rest mass above the ground state energy; adding a similarly small amount of energy (13.6 eV) can release the electron. The neutron does not have any nearby excited states (in the same sense) and cannot be split into a proton and an electron. A hydrogen atom is about a bohr radius in size (10^{-10} m) while a neutron is much smaller, approximately the same size as the proton or nucleus of hydrogen (10^{-14} m), which is much smaller.

Question. Of the four known forces of nature, which of them can affect (a) electrons, (b) neutrinos, and (c) protons?

Answer. (a) Electrons are affected by the electromagnetic force (they have charge), the weak force (they have weak charge) and gravitation (they have mass). (b) Neutrinos without electric charge or mass are affected only by the weak force. (c) Protons, with electromagnetic charge, weak charge, strong charge, and mass are affected by all four forces.

EXAMPLES AND SOLUTIONS

Example 1

The Z_0 particle, one of the neutral mediators of the weak interaction, has a mass of 91 GeV/c². It can be produced at rest in a high energy electron-positron collider in which electrons and positrons in counter rotating circular beams are made to collide at intersection points. Estimate the product of the design parameters, rB, where r is the radius of the circulating beams and B is the magnetic field strength holding the electrons and positrons in their circular path.

Solution:

The electrons and positrons of energy E annihilate to produce a Z_0 at rest:

$$2E = m_{Z_0}c^2$$

$$E = \frac{91}{2} \text{ GeV} = 45.5 \text{ GeV}$$

The electrons and positrons, with mass m_e, have momentum p that can be calculated from:

$$E^2 = p^2c^2 + m_e^2c^4$$

$$p = \frac{1}{c}\sqrt{\left(E^2 - m_e^2c^4\right)}$$

However the electron mass energy ($m_e c^2$) is about 0.5 MeV << E so that to a good approximation

$$p \cong \frac{E}{c} = 45.5 \frac{\text{GeV}}{c}$$

In a magnetic field of strength B, a non-relativistic particle has a radius r where

$$r = \frac{mv}{qB} = \frac{p}{qB}$$

where the expression r = p/qB is also correct at relativistic speeds. Thus

$$rB = \frac{p}{q} = \frac{45.5 \text{ GeV}/c}{e}$$

We convert units using

$$1 \text{ eV} = 1.6 \times 10^{-19} \text{ J}$$

$$1 \text{ GeV} = 10^9 \text{ eV}$$

so

$$rB = \frac{(45.5 \times 10^9 \text{ eV})(1.6 \times 10^{-19} \text{ J/eV})}{(1.6 \times 10^{-19} \text{ C})(3 \times 10^8 \text{ m/s})}$$

$$rB = 151.7 \text{ m·Tesla} \approx 150 \text{ m·Tesla}$$

With magnets of strength 0.1 Tesla, the radius of ring needed is 1.5 km. At the e+e- collider in Geneva, Switzerland, weaker magnets are used (requiring a larger machine) to reduce the synchrotron radiation.

Example 2

A neutral rho (ρ^0) meson (mass 770 MeV/c2) may decay at rest into a pair of charged pions (mass 140 MeV/c2)

$$\rho^0 \rightarrow \pi^+ + \pi^-$$

or into an electron and positron pair

$$\rho^0 \rightarrow e^+ + e^-$$

Find the energies and momenta of the pions and electrons.

Solution:

If M is the ρ^0 meson mass and m the decay product mass, the conservation of energy requires that

$$Mc^2 = 2E = 2\sqrt{p^2 c^2 + m^2 c^4}$$

when p is the momentum of the decay product particles, equal for each in the pair because momentum is also conserved; the ρ^+ and ρ^- momenta are equal and opposite to produce zero net final momentum, equal to the initial momentum of zero (the rho is at rest).

Then the energies of the decay products are

$$E = \sqrt{p^2 c^2 + m^2 c^4} = \frac{Mc^2}{2} = 385 \text{ MeV}$$
$$E = E_{\pi^+} = E_{\pi^-} = E_{e^+} = E_{e^-}$$

The momenta, however are different,

$$p = \frac{1}{c}\sqrt{(E^2 - m^2 c^4)}$$

since they depend on the decay particle mass

$$p_{\pi^{\pm}} = \frac{1}{c}\sqrt{\left(E^2 - m_{\pi}^2 c^4\right)}$$

Since $m_p c^2 = 140$ MeV,

$$p_{\pi^{\pm}} = \frac{1}{c}\sqrt{\left((385)^2 - (140)^2\right)} \approx 358 \text{ MeV/c}$$

For electrons and positrons,

$$p_{e^{\pm}} = \frac{1}{c}\sqrt{\left(E^2 - m_e^2 c^4\right)}$$

$$p_{e^{\pm}} = \frac{1}{c}\sqrt{\left((385)^2 - (0.5)^2\right)} \approx 385 \text{ MeV/c}$$

Example 3

Protons in a cyclotron spiral out to a radius of 15 cm. The magnetic field has a magnitude of 1.25 T. (a) Find the frequency of the alternating voltage used to accelerate the protons in the gap. (b) Find the energy of the protons, in MeV.

Solution:

(a) The cyclotron frequency, or angular frequency of rotation of the protons

$$\omega = \frac{Be}{m} = \frac{(1.25 \text{ T})(1.6 \times 10^{-19} \text{ C})}{(1.6 \times 10^{-27} \text{ kg})}$$

$$= 1.2 \times 10^8 \text{ rad/s}.$$

The frequency of the alternating voltage is

$$f = \frac{\omega}{2\pi} = 1.91 \times 10^7 \text{ Hz}$$

(b) The kinetic energy of the protons is

$$\frac{1}{2} mv^2 = \frac{1}{2} m(\omega r)^2$$

$$\frac{1}{2} m(\omega r)^2 = \frac{1}{2} (1.6 \times 10^{-27} \text{ kg})\left[(1.2 \times 10^8 \text{ s}^{-1})(0.15 \text{ m})\right]^2$$

$$\frac{1}{2} m(\omega r)^2 = \frac{2.6 \times 10^{-13} \text{ J}}{1.6 \times 10^{-19} \text{ J/eV}} = 1.6 \text{ MeV}$$

Example 4

Consider the following potential decays of the positively charged pion:

(1) $\pi^+ \rightarrow \mu^+ \nu_\mu$
(2) $\pi^+ \rightarrow e^+ \nu_e$
(3) $\pi^+ \rightarrow \mu^+ \nu_\mu \gamma$
(4) $\pi^+ \rightarrow e^+ \nu_e \gamma$
(5) $\pi^+ \rightarrow e^+ \nu_e \pi^0$
(6) $\pi^+ \rightarrow e^+ \nu_e e^+ e^-$
(7) $\pi^+ \rightarrow e^+ \nu_e \nu \overline{\nu}$
(8) $\pi^+ \rightarrow \mu^+ \overline{\nu}_e$
(9) $\pi^+ \rightarrow \mu^+ \nu_e$
(10) $\pi^+ \rightarrow \mu^- e^+ e^- \nu$

Which ones are allowed by lepton number conservation?

Solution:

Referring to Table 44-2 of the text, the lepton number balance in the various decays can be tabulated as follows:

(1)	$0 \rightarrow$ -1 + 1	(muon number)
(2)	$0 \rightarrow$ -1 + 1	(electron number)
(3)	$0 \rightarrow$ -1 + 1 + 0	(muon number)
(4)	$0 \rightarrow$ -1 + 1 + 0	(electron number)
(5)	$0 \rightarrow$ -1 + 1 + 0	(electron number)
(6)	$0 \rightarrow$ -1 + 1 + (-1) + 1	(electron number)
(7)	$0 \rightarrow$ -1 + 1 + 1 + (-1)	(electron number)
(8)	$0 \rightarrow$ -1 - 1	(lepton number)
(9)	$0 \rightarrow$ -1 + 1	(mixed lepton number)
(10)	$0 \rightarrow$ 1 + (-1) + (-1) + 1	(mixed lepton number)

Decays (1) through (7) are allowed by strict conservation of muon number and electron number separately. Decay (8) violates lepton number even if only the sum of muon and electron number were conserved. Decays (9) and (10) conserve the sum of electron and muon number but violate individual electron and muon number conservation.

Decays (1) through (6) are actually seen. Decay (7) has not been seen, nor have the lepton violating decays (8) through (10) been seen.

Example 5

Find the charge and strangeness of all mesons that can be constructed from quark-antiquark pairs with "flavors" u, d, and s.

Solution:

There are 3 ways of choosing the quark and 3 ways of choosing the antiquark or 9 independent combinations:

$$u\,\bar{u} \qquad\qquad u\,\bar{d} \qquad\qquad u\,\bar{s}$$

$$d\,\bar{u} \qquad\qquad d\,\bar{d} \qquad\qquad d\,\bar{s}$$

$$s\,\bar{u} \qquad\qquad s\,\bar{d} \qquad\qquad s\,\bar{s}$$

Three of these have no strangeness or electric charge; $S = 0$ and $Q = 0$. They are: $u\,\bar{u}$, $d\,\bar{d}$, and $s\,\bar{s}$.

The others are charged and have the quantum numbers listed below;

$q\,\bar{q}$	Q	S	Seen in nature as: (Spin 0, negative parity multiplet)
$u\,\bar{d}$	+ 1	0	$\pi+$
$d\,\bar{u}$	- 1	0	$\pi-$
$u\,\bar{s}$	+ 1	+ 1	K+
$s\,\bar{u}$	- 1	- 1	K-
$s\,\bar{d}$	0	- 1	K^0
$d\,\bar{s}$	0	+ 1	K^0

Example 6

According to Hubble's law, matter at a distance r travels away from us at a speed

$$v = H_0 r$$

where H_0 is Hubble's constant,

$$H_0 = 17 \times 10^{-3} \text{ m/s per lightyear.}$$

Assuming a Big Bang scenario, what is the age of the universe?

Solution:

If all the matter originated from a point, it would take a time T to run the universe "backward" to that point (in time)

$$T = \frac{r}{v} = \frac{r}{H_0 r} = \frac{1}{H_0}$$

$$T = \left[\frac{17 \times 10^{-3} \text{ m/s}}{\left(3 \times 10^8 \text{ m/s}\right) \text{year}} \right]^{-1} = 1.76 \times 10^{10} \text{ years}$$

In this picture, we are at that "point of creation" but any other observer can make the same imaginary trip backward in time and arrive at the same big bang.

QUIZ

1. In a formerly proposed accelerator (the Superconducting Super Collider, or SSC) protons were intended to achieve momenta of 20 TeV/c = 20 x 10^{12} eV/c, where c is the velocity of light. If the bending magnets that were to keep the protons moving in a circle had a magnetic induction of 6.6 Tesla, estimate the radius of curvature of the protons in the magnetic field.

Answer: 10 km

2. A proton (charge +e) is known to be composed of two "up" quarks and one "down" quark. A neutron (charge zero) is known to be composed of one "up" and two "down" quarks. From this information alone, calculate the charge on the up and down quarks.

Answer: q_u = 2e/3, q_d = -e/3

3. Which of the following decays conserve lepton number?

(a) $\pi^0 \to 2\gamma$
(b) $\pi^0 \to \mu^+ e^-$
(c) $\mu^- \to e^- \overline{\nu_e} \nu_\mu \gamma$
(d) $\mu^- \to e^- \nu_e \overline{\nu_\mu}$
(e) $\mu^- \to e^- \gamma\gamma$
(f) $\mu^- \to e^- \overline{\nu_e} \nu_\mu$

Answer: (a), (c), and (f)

4. The unstable particle Φ with mass 1020 MeV/c^2 decays into a pair of charged K mesons (mass 494 MeV/c^2 each).

$\Phi \to K^+ K^-$

What is the momentum of the K+ in the Φ rest system?

Answer: 127 MeV/c